53
Topics in Heterocyclic Chemistry

Aims and Scope

The series Topics in Heterocyclic Chemistry presents critical reviews on present and future trends in the research of heterocyclic compounds. Overall the scope is to cover topics dealing with all areas within heterocyclic chemistry, both experimental and theoretical, of interest to the general heterocyclic chemistry community.

The series consists of topic related volumes edited by renowned editors with contributions of experts in the field. All chapters from Topics in Heterocyclic Chemistry are published OnlineFirst with an individual DOI. In references, Topics in Heterocyclic Chemistry is abbreviated as Top Heterocycl Chem and cited as a journal.

More information about this series at http://www.springer.com/series/7081

Oleg V. Larionov
Editor

Heterocyclic *N*-Oxides

With contributions by

D.E. Chavez · M. Kotora · P. Koukal · O.V. Larionov ·
R. Loska · D. Nečas · J.S. Poole · D.E. Stephens · J. Ulč

Springer

Editor
Oleg V. Larionov
University of Texas at San Antonio
San Antonio, Texas
USA

ISSN 1861-9282 ISSN 1861-9290 (electronic)
Topics in Heterocyclic Chemistry
ISBN 978-3-319-60686-6 ISBN 978-3-319-60687-3 (eBook)
DOI 10.1007/978-3-319-60687-3

Library of Congress Control Number: 2017945375

Printed on acid-free paper

This Springer imprint is published by Springer Nature
The registered company is Springer International Publishing AG
The registered company address is: Gewerbestrasse 11, 6330 Cham, Switzerland

Preface

Heterocyclic *N*-oxides have gained in prominence in many areas of chemistry in the past several decades. In the area of organic synthesis, *N*-oxides have emerged as important substrates for regioselective functionalization of C–H bonds and cycloaddition reactions. There has also been a surge in interest in the catalytic, energetic, and photochemical properties of *N*-oxides. This volume seeks to provide an update on the recent advances in these important areas of chemistry of heterocyclic *N*-oxides. In the first chapter David E. Chavez gives an in-depth overview of the progress in the studies that aim to exploit the unique structural and electronic properties of *N*-oxides for the development of novel energetic materials (see chapter "Energetic Heterocyclic *N*-Oxides"). The moderate Lewis basicity of the oxygen atom in *N*-oxides has been employed in the design of catalysts for a variety of asymmetric transformations. Martin Kotora et al. examine the current state of the art in catalytic applications of heterocyclic *N*-oxides (see chapter "Pyridine *N*-Oxides and Derivatives Thereof in Organocatalysis"). *N*-Oxide functionality has emerged as a versatile directing group in the burgeoning field of C–H functionalization of N-heterocycles. David E. Stephens and Oleg V. Larionov survey recent advances in transition metal-catalyzed C–H functionalization of azine and azole *N*-oxides with the focus on transformations that retain the *N*-oxide functionality (see chapter "Transition Metal-Catalyzed C–H Functionalization of Heterocyclic *N*-Oxides"). Cycloaddition reactions of heterocyclic *N*-oxides play an important role in the synthesis of nitrogen-containing heterocycles. Rafał Loska discusses mechanisms and synthetic applications of cycloaddition reactions of azine and azole *N*-oxides (see chapter "Recent Advances in Cycloaddition Reactions of Heterocyclic *N*-Oxides"). The photoinduced transformations of heteroarene *N*-oxides have been intensively studied since the early days of heterocyclic chemistry. An excellent overview of the current status of photochemistry of *N*-oxides is given by James S. Poole (see chapter "Recent Advances in the Photochemistry of Heterocyclic *N*-Oxides and their Derivatives").

The aim of this book is to shed light on some of the most exciting developments in the chemistry of heterocyclic *N*-oxides and to demonstrate the versatility of their applications across a wide range of fields – from energetic materials to catalysis, and from photochemistry to organic synthesis.

I thank the authors and the editorial staff at Springer for their dedication and efforts that have led to the production of this book.

San Antonio, TX, USA
February 2017

Oleg V. Larionov

Contents

Contents

Top Heterocycl Chem (2017) 53: 1–28
DOI: 10.1007/7081_2017_5

Published online: 20 April 2017

Energetic Heterocyclic *N*-Oxides

David E. Chavez

Abstract Heterocyclic *N*-oxides have been shown to be useful in energetic materials applications such as propellants, explosives, and pyrotechnics. This chapter provides a survey of a variety of different heterocyclic *N*-oxides that have been studied for their energetic materials properties. Where possible, information such as heat of formation, density, detonation pressure (P_{CJ}), and detonation velocity (V_D) are provided.

Keywords 1,2,3-Triazole • 1,2,4-Triazole • Energetic materials • Explosives • Furoxan • Hypofluorous acid • Macrocycles • Oxone • Pyrazine • Pyrazole • Pyridine • Tetrazine • Tetrazole • Triazine

Contents

D.E. Chavez (✉)
Los Alamos National Laboratory, Los Alamos, NM 87544, USA
e-mail: dechavez@lanl.gov

1 Introduction

Heterocyclic *N*-oxides have received intense interest in the field of energetic materials in the past decade. This is in large part due to the fact that the introduction of an *N*-oxide functional group on to a heterocyclic system is a strategy that attempts to make use of the interesting properties *N*-oxide materials tend to display. The zwitterionic nature of the *N*-oxide bond tends to lead to large dipole moments, which often leads to increases in crystal densities, a property that is highly important with respect to energetic material performance. *N*-oxides also increase the oxygen balance of a molecule, providing more oxidizer to improve the overall combustion of the fuel in the molecule. Additionally, *N*-oxides are also known to provide a stabilizing effect for some heterocyclic systems. Introduction of the *N*-oxide functional group can also lead to reduction in the sensitivity of an energetic material toward destructive stimuli, such as impact, spark, or friction.

The strategy used to introduce *N*-oxides onto the majority of the heterocycles discussed in this section (with the exception of furoxans) is through oxidation of ring nitrogen atoms. One of the most often used oxidants in this regard is Oxone. Other peracids are used as well, including peracetic acid and peroxytrifluoroacetic acid. For particularly unreactive heterocyclic systems, hypofluorous acid has also been employed. Examples of the formation of heterocyclic *N*-oxides directly through cyclization or condensation reactions are seen less frequently but are common in the formation of furoxans and some tetrazole and triazole *N*-oxides.

This chapter is intended to provide an overview of novel, energetic heterocyclic *N*-oxides that have been prepared over the past 15–20 years, though there are a few examples from earlier work that have been included. This chapter is not intended to be a comprehensive review of all of the heterocyclic *N*-oxide energetic material work that has been reported, but rather, it is a survey that highlights energetic, heterocyclic *N*-oxide syntheses and characterizations. This chapter attempts to discuss *N*-oxide derivative of five-membered heterocycles (pyrazoles, triazoles, tetrazoles, furoxans) and six-membered heterocycles (pyridines, pyrazines, triazines, and tetrazines) as well as some fused combinations of different heterocycles while providing energetic materials properties where the data exists.

2 Pyrazole Oxides

2.1 *Dinitropyrazole 1-Oxides*

Shevelev reported the synthesis of *N*-hydroxydinitropyrazoles in 1996 [1]. Both 3,4- and 3,5-dinitropyrazoles were oxidized to the corresponding 1-hydroxy compounds using a buffered $KHSO_5$ solution (Scheme 1). The yields were modest (20–48%). No salts were prepared nor were any explosive properties of these materials reported. While these materials were not converted to *N*-oxide salts by reaction with bases, the work opened up the possibility for other efforts to proceed along those lines and, thus, is an important contribution to energetic heterocyclic *N*-oxide chemistry.

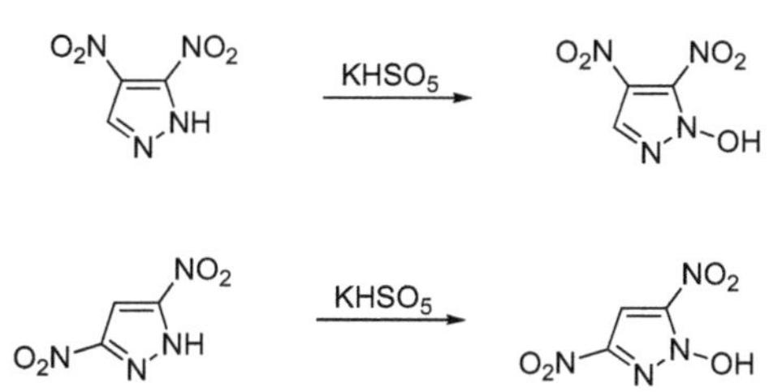

Scheme 1 The synthesis of *N*-hydroxydinitropyrazoles

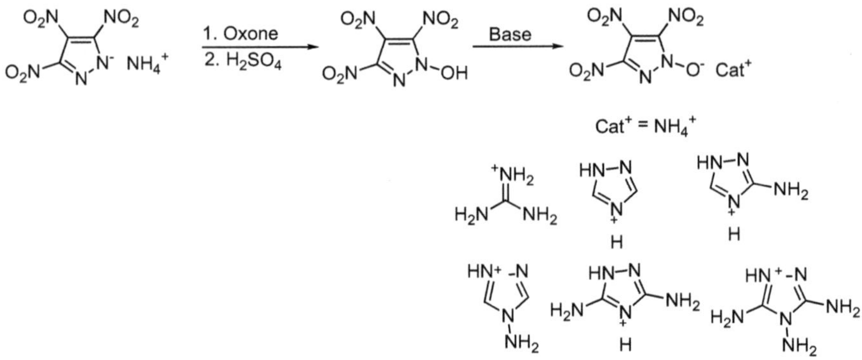

Scheme 2 The synthesis of some salts of 3,4,5-trinitropyrazole 1-oxides

2.2 3,4,5-Trinitropyrazole 1-Oxide

Shreeve reported the oxidation of 3,4,5-trinitropyrazole with Oxone in 2012 [2]. Reaction of the ammonium salt of 3,4,5-trinitropyrazole proceeded at 55°C for 3 days to give the 1-oxide (Scheme 2). Acidification provided the free acid, which was then converted to a variety of nitrogenous salts. The most promising salt was the ammonium salt. This material has a density of 1.82 g/cm^3, has a heat of formation of 118 kJ/mol, and is thermally stable up to 176°C. It has an excellent oxygen balance and has impact sensitivity similar to conventional high explosives such as RDX (1,3,5-trinitroperhydro-1,3,5-triazine). The explosive performance was predicted to be good ($V_D = 8{,}676$ km/s, $P_{CJ} = 35$ GPa).

3 Triazole Oxides

3.1 1,2,3-Triazole 1-Oxides

Furoxans have been shown to react with alkylamines to form triazole 1-oxides [3]. In their work toward the synthesis of tetrazino-1,2,3,4-tetrazine dioxide, Churakov and coworkers found that a variety of alkylamines reacted with 3-*tert*-butylazoxy-4-aminofuroxan to give the corresponding 1,2,3-triazole 1-oxide (Scheme 3). Treatment of the *tert*-butyl derivative with trifluoroacetic acid or gaseous HCl easily removes the *tert*-butyl group.

Direct oxidation of the 1,2,3-triazole was reported for 4,5-dinitro-1,2,3-triazole, which can be oxidized to the 1-oxide derivative with Oxone in the presence of K_2HPO_4 [4]. After 48 h at 55°C, the free acid of 4,5-dinitro-1,2,3-triazole 1-oxide is formed with 96% conversion (Scheme 4). The acid can be converted to the potassium or ammonium salts with potassium carbonate or ammonia. The crystal

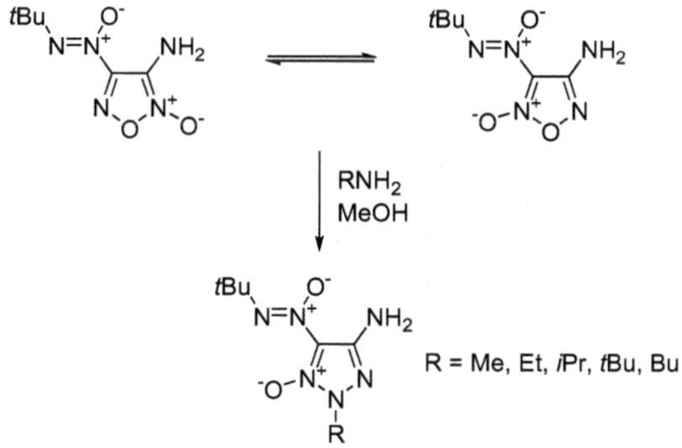

Scheme 3 The synthesis of some 1,2,3-triazole 1-oxides

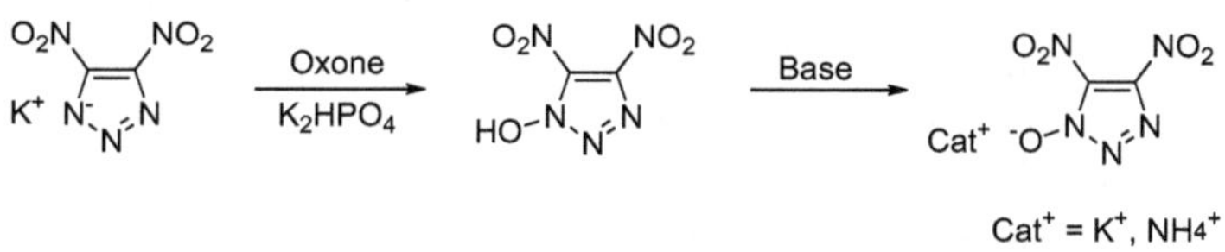

Scheme 4 The synthesis of potassium and ammonium salts of 4,5-dinitro-1,2,3-triazole 1-oxide

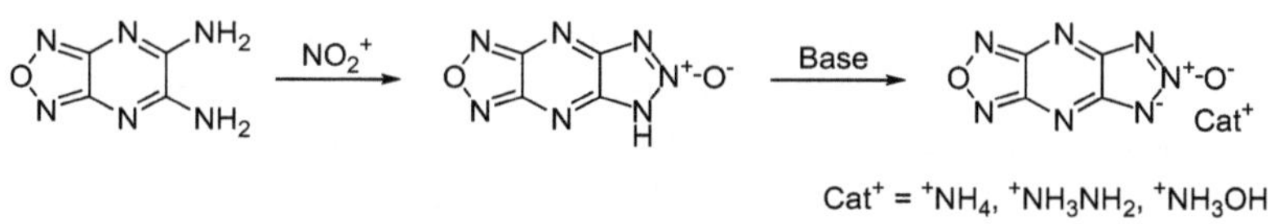

Scheme 5 The synthesis of 1,2,3-triazolo[4,5-*e*]furazano[3,4-*b*]pyrazine 6-oxide and some salts

density of the ammonium salt was found to be 1.789 g/cm^3 and this material was found to be thermally stable up to 195°C.

Treatment of 5,6-diaminofurazano[3,4-*b*]pyrazine with nitronium ion results in a condensation and cyclization reaction to form 1,2,3-triazolo[4,5-*e*]furazano[3,4-b] pyrazine 6-oxide (Scheme 5) [5]. The N–H group in this molecule is acidic enough to react with nitrogenous bases, as Shreeve showed in 2014 [6]. The ammonium, hydrazinium, and hydroxylammonium salts were prepared along with some salts with heterocyclic cations. The materials ranged in thermal stability from 141°C for the hydroxylammonium salt to 281°C for the free acid. The free acid also showed the highest density (1.85 g/cm^3) and the most promising calculated performance ($V_D = 8.53$ km/s, $P_{CJ} = 32.4$ GPa). All of the materials were relatively insensitive to impact (30–40 J).

3.2 1,2,4-Triazole 1-Oxides

Oxidation of 3-nitro-1,2,4-triazole was briefly mentioned in a publication by Shevelev where the oxidation of pyrazoles had been studied [1]. $KHSO_5$ was the oxidant found to be the most successful for the 3-nitro-1,2,4-triazole substrate, providing the 1-*N*-hydroxy derivative. No energetic materials properties for the product were reported, nor were any salts prepared.

A much more challenging triazole to oxidize is 3,5-dinitro-1,2,4-triazole, but a method was reported in 2002 [7]. The ultimate goal of the research was to prepare highly energetic liquid propulsion ingredients. Here again, Oxone was used as the oxidant, but the reaction was found to be slow and inefficient, and pure salts of 3,5-dinitro-1,2,4-triazole 1-oxide were not isolated. In 2012, a different oxidant was used to prepare pure materials [4]. Hypofluorous acid oxidation proceeded within 15 min giving a 65% yield of the potassium salt of 3,5-dinitro-1,2,4-triazole 1-oxide (density = 2.08 g/cm^3, T_{dec} = 215°C) (Scheme 6). The ammonium salt had a crystal density of 1.784 g/cm^3 and a heat of formation of 25 kcal/mol. The material begins to decompose at 145°C.

Klapoetke and coworkers prepared several energetic salts of 3,3′-dinitrobis (1,2,4-triazole) 1,1′-dioxide [8]. 3,3′-Dinitrobis(1,2,4-triazole) was oxidized with Oxone in the presence of a potassium acetate buffer. When the reaction was complete, the reaction mixture was treated with sulfuric acid and extracted to give the diol. The diol was then treated with a variety of bases (ammonia, hydrazine, hydroxylamine, guanidine, aminoguanidine, triaminoguanidine) (Scheme 7). Oxidation was found to occur only at the 1 and 1′ positions. The diol was found to be thermally stable to 191°C and had gas pycnometry density of 1.91 g/cm^3. Conversion to the nitrogenous salts improves the thermal stability in all cases

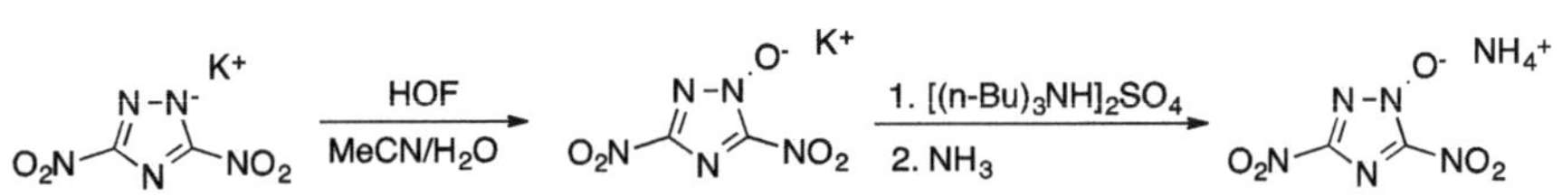

Scheme 6 The synthesis of the potassium and ammonium salts of 3,5-dinitro-1,2,4-triazole 1-oxide

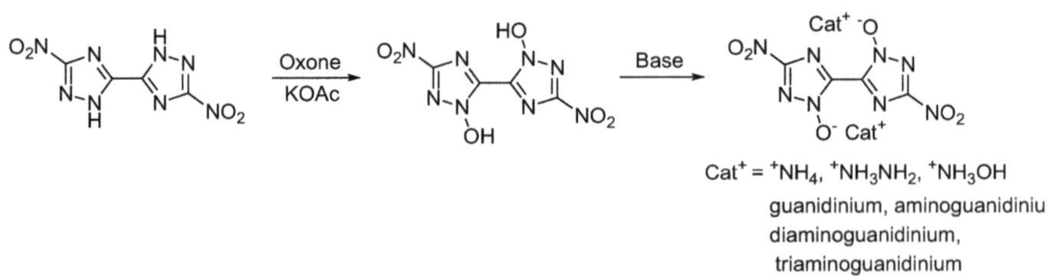

Scheme 7 The synthesis of energetic salts of 3,3′-dinitrobis(1,2,4-triazole) 1,1′-dioxide

(207–329°C), with the guanidinium salt being the most thermally stable. In general, the densities of the salts are decreased compared to the parent diol compound, with the hydroxylamine salt having the highest density (1.90 g/cm^3). All of the salts have low sensitivity to impact (>40–15 J) and are essentially insensitive to friction. The salt with the most promising explosive properties was the hydroxylamine salt, with a predicted detonation pressure of 39 GPa and a detonation velocity of 9.1 km/s, comparable to high-performing conventional explosives such as HMX (octahydro-1,3,5,7-tetranitro-1,3,5,7-tetrazocine). No toxicity studies were reported for any of the compounds.

4 Tetrazole Oxides

4.1 5-Nitrotetrazole 2-Oxide

5-Nitrotetrazoles can be oxidized to the corresponding 2-*N*-oxide with Oxone, in the presence of an acetate buffer (Scheme 8) [7]. No 1-oxide isomer was observed under these reaction conditions. Various salts can be prepared using ion exchange resins. The hydroxylamine salt was reported to have an experimental heat of formation of 40 kcal/mol and a density of 1.85 g/cm^3. The explosive performance properties of the energetic salts were reported by Klapoetke [9]. The hydroxylamine salt was predicted to have a detonation pressure of 39 GPa and a detonation velocity of 9.45 km/s. The other salts displayed lower performance. Thermal studies of the hydroxylamine amine salt showed it began to decompose at 157°C at a 5°C heating ramp rate. It was also relatively sensitive to impact (4 J) and friction (60 N). Interestingly, all of the 2-*N*-oxide salts studied were less sensitive than the corresponding 5-nitrotetrazolate salts, with the exception of the aminoguanidinium salt.

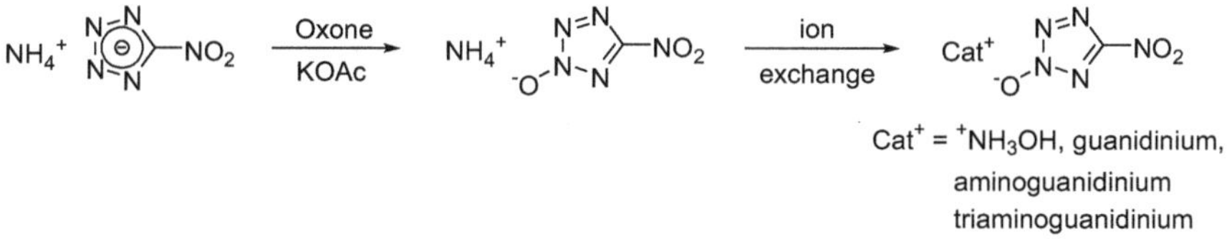

Scheme 8 The synthesis of 5-nitrotetrazole 2-oxide and some energetic salts

4.2 5-Azidotetrazole 2-Oxide

In 2011, the Klapoetke group attempted to study the oxidation of the extremely sensitive compound 5-azidotetrazole [10]. This compound is quite hazardous to work with and can be considered a contact explosive. 5-Azidotetrazole was oxidized with Oxone in the presence of potassium acetate as a buffer. The reaction proceeded at 40°C for 3 days (Scheme 9). Acidification, extraction with ether, and treatment with ammonia provided the ammonium salt. The silver, potassium, and sodium salts were prepared along with the aminoguanidinium salt. All of the salts displayed lower thermal stability than the corresponding 5-azidotetrazole salts. Here again the sensitivity to impact and friction was less for the 2-*N*-oxide derivatives compared to respective non-oxide compounds. The density of the ammonium salt was 1.689 g/cm^3 and the detonation velocity and pressure were calculated to be 8.92 km/s and 32.5 GPa, respectively.

4.3 5-Aminotetrazole 2-Oxide

An interesting tetrazole *N*-oxide synthesized by the Klapoetke group is 5-aminotetrazole 1-oxide [11]. Rather than accessing this heterocycle through oxidation of 5-aminotetrazole, it was prepared by reacting cyanogen azide with hydroxylamine. In the presence of excess hydroxylamine, the hydroxylammonium salt can be isolated (Scheme 10). Treatment with HCl, to protonate the oxide, followed by treatment with ammonia provides the ammonium salt. The hydroxylamine salt begins to decompose at 155°C and was shown to have two polymorphs, one with a density of 1.664 g/cm^3 and the other with 1.735 g/cm^3. The ammonium salt had a much lower density (1.53 g/cm^3) and an onset of decomposition of 195°C. These materials were relatively insensitive to impact (>40–10 J) and insensitive to friction (>360 N). The high-density hydroxylammonium salt showed the

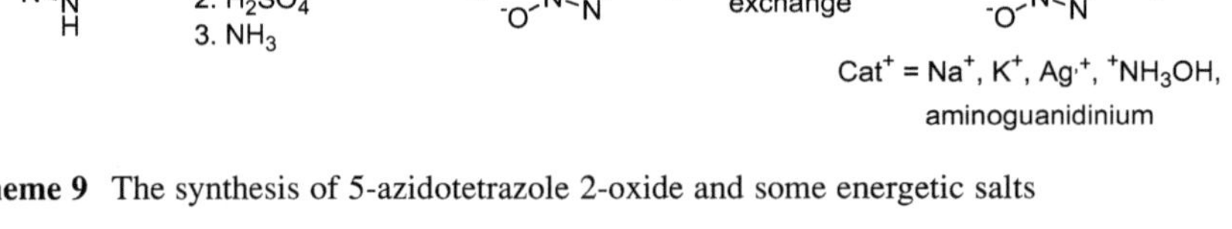

Scheme 9 The synthesis of 5-azidotetrazole 2-oxide and some energetic salts

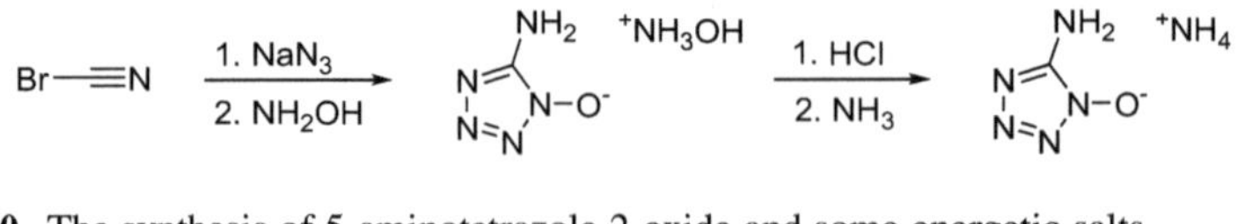

Scheme 10 The synthesis of 5-aminotetrazole 2-oxide and some energetic salts

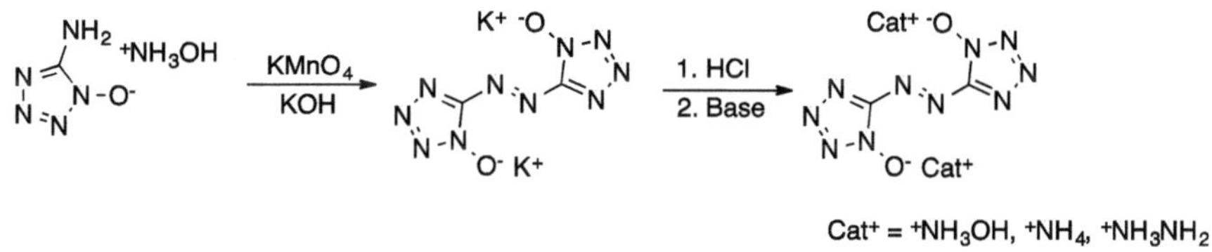

Scheme 11 The synthesis of salts of 5,5′-azobistetrazole 1,1′-dioxide

most promise with respect to explosive performance (P_{CJ} = 35.7 GPa, V_D = 9.3 km/s).

Amine oxidation of 5-aminotetrazole 1-oxide is possible using basic potassium permanganate resulting in the bis-potassium salt of the azo-coupled product (Scheme 11) [11]. Treatment of the potassium salt with acid followed by reaction with nitrogen bases gives access to the hydroxylammonium, ammonium, and hydrazinium salts. The nitrogenous salts displayed impact sensitivities in the range of 3–15 J and friction sensitivities in the range of 20–160 N. The densities ranged from 1.725 g/cm^3 for the hydrazinium salt to 1.80 g/cm^3 for the hydroxylammonium salt. Amazingly the ammonium salt was quite thermally stable (decomposition onset 250°C), while the other salts decomposed at about 190°C. The best performer was the hydroxylammonium salt (P_{CJ} = 37.5 GPa, V_D = 9.35 km/s).

4.4 *Dihydroxylammonium 5,5′-Bistetrazole 1,1′-Dioxide*

One of the most promising energetic *N*-oxide heterocycles discovered recently is the dihydroxylammonium salt of 5,5′-bistetrazole 1,1′-dioxide (TKX-50) [12]. TKX-50 is described as a molecule that is easily prepared and very powerful but with the required thermal stability, low toxicity, and safety properties to be used as an RDX replacement. Two different approaches were used to access the TKX-50 structure. The first was oxidation of 5,5′-bitetrazole using potassium acetate-buffered Oxone. Unfortunately, the method resulted in the production of both the 1,1′-di-*N*-oxide (11%) and the 2,2′-di-*N*-oxide (58%) (Scheme 12). A different route to TKX-50 is based upon the work of Tselinksi [13]. This method involves the reaction of sodium azide with dichloroglyoxime, followed by treatment with HCl in ether to access the TKX-50 precursor diol.

TXK-50 was found to have an impact sensitivity of 20 J and a friction sensitivity of 120 N. These values compare very favorably with conventional energetic materials such as RDX (impact = 7.5 J, friction = 120 N). It is also thermally stable up to 221°C; TKX-50 has a density of 1.877 g/cm^3 at 298 K and a heat of formation calculated to be 446.6 kJ/mol. These data taken together were used to predict the explosive performance of TKX-50. The detonation velocity was predicted to be 9.7 km/s and a detonation pressure of 42.4 GPa. These data were compared to other standard explosives to show that TKX-50 was predicted to

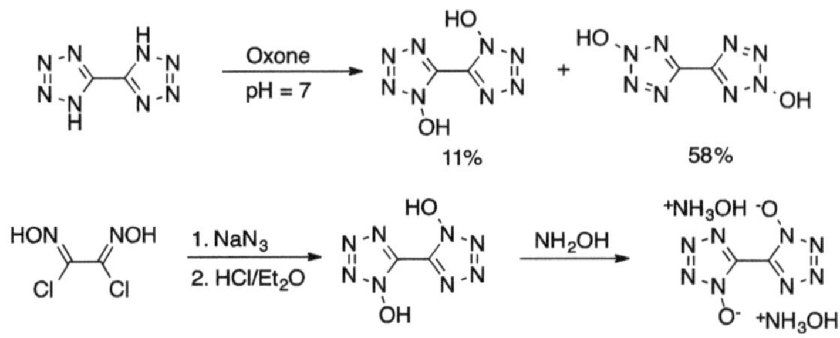

Scheme 12 Synthesis routes to TKX-50

outperform most other conventional explosives with respect to detonation velocity and was only surpassed by CL-20 with respect to detonation pressure. Toxicity studies were also performed on *Vibrio fischeri*, and the results showed that TKX-50 was less toxic to this microorganism than RDX.

4.5 Amino-Hydroximoyl-Tetrazole 2-Oxides

Another example of the use of the *N*-oxide to improve the performance of tetrazole compounds is seen in the synthesis of amino-hydroximoyl-tetrazole 2-oxides [14]. The *N*-oxide was installed prior to the formation of the aminohydroximoyl group by oxidizing cyanotetrazole, which provided cyanotetrazole-2-oxide (Scheme 13). Subsequent treatment with hydroxylamine provided the amino-hydroximoyl-tetrazole-2-*N*-hydoxy compound. Treatment with nitrogenous bases allows access to a variety of salts (hydroxylammonium, guanidinium, amino-guanidinium, ammonium, triaminoguanidinium). In the report, the hydroxyl-ammonium salt was compared to the hydroxylammonium salt of the unoxidized amino-hydroximoyl-tetrazole. Here it was shown that the *N*-oxide improves the insensitivity toward impact (>40 vs 10 J), while the thermal stability is reduced slightly ($T_{dec} = 164$ vs 171°C). A slight decrease in the heat of formation (282 vs 294 kJ/mol) and a significant increase in density (1.704 vs 1.639 g/cm^3) lead to an increase in calculated detonation velocity by 3.6% (8.934 vs 8.643 km/s) and an increase in calculated detonation pressure by 12% (30.9 vs 27.2 GPa).

5 Furoxans

5.1 4,4′-Dinitro-3,3′-Diazofuroxan

A very high-energy furoxan was reported by Binnikov in 1999, namely, 4,4′-dinitro-3,3′-diazofuroxan [15]. This molecule was synthesized by azo coupling of 4-amino-3-

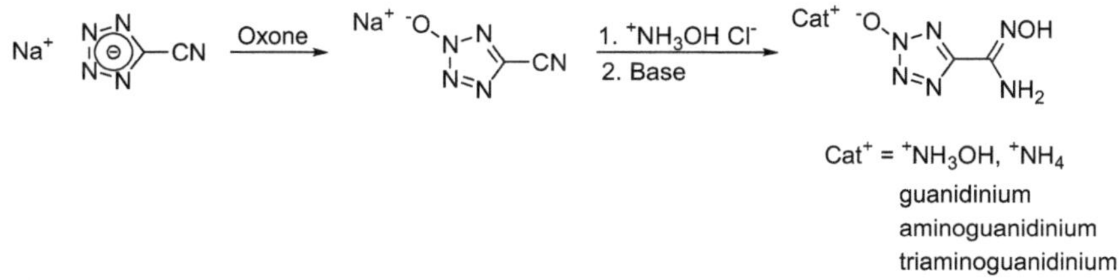

Scheme 13 Synthesis of salts of amino-hydroximoyl-tetrazole 2-oxide

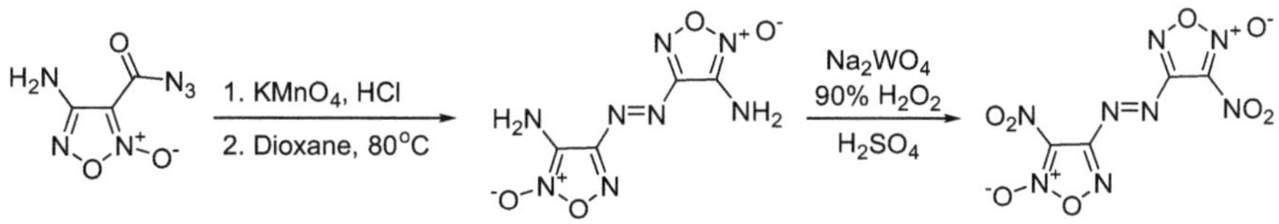

Scheme 14 Synthesis of 4,4′-dinitro-3,3′-diazofuroxan

azidocarbonylfuroxan. A Curtius rearrangement was used to access the corresponding 4,4′-diamino compound. Oxidation using Na_2WO_4 in the presence of sulfuric acid and 90% peroxide provided 4,4′-dinitro-3,3′-diazofuroxan (Scheme 14). An experimental detonation velocity was determined to be 10 km/s, with a crystal density of 2.02 g/cm^3. This is one of the fastest detonation velocities ever reported. The high density is due in large part to very efficient crystal packing.

5.2 *Dinitrofurazanofuroxan*

3,4-bis(3-nitrofurazan-4-yl)furoxan (BNFF) is a promising furoxan derivative that displays good performance and physical properties. The first appearance of BNFF in the literature was in a report by Sheremetev and coworkers [16]. Subsequently, another report was published, with few synthesis details [17]. In 2010, Chung published a detailed synthesis procedure for the preparation of BNFF, a portion of which is included in Scheme 15. The synthesis includes diazotization of the amide oxime compound in the presence of HCl to provide the oximyl chloride, treatment of the oximyl chloride with base to produce a nitrile oxide intermediate that dimerizes to provide 3,4-bis(3-aminofurazan-4-yl)furoxan (BAFF). Oxidation with peroxytrifluoroacetic acid (PTFA) provided BNFF.

BAFF can be crystallized from water or ethanol, which results in two different polymorphs with densities of 1.745 and 1.737 g/cm^3, respectively. The high-density material has a predicted detonation velocity of 8.1 km/s and an experimental detonation velocity of 7.18 km/s at a density of 1.53 g/cm^3. It was reported to be insensitive to impact, spark, and friction [18].

Scheme 15 Synthesis of 3,4-bis(3-aminofurazan-4-yl)furoxan (BAFF) and 3,4-bis(3-nitrofurazan-4-yl) furoxan (BNFF)

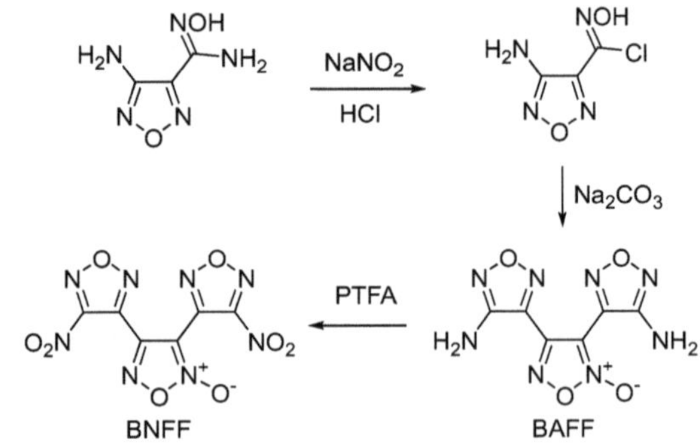

Scheme 16 Synthesis of 3,4-bis(3-nitrofurazan-4-yl) furoxan (DAZTF)

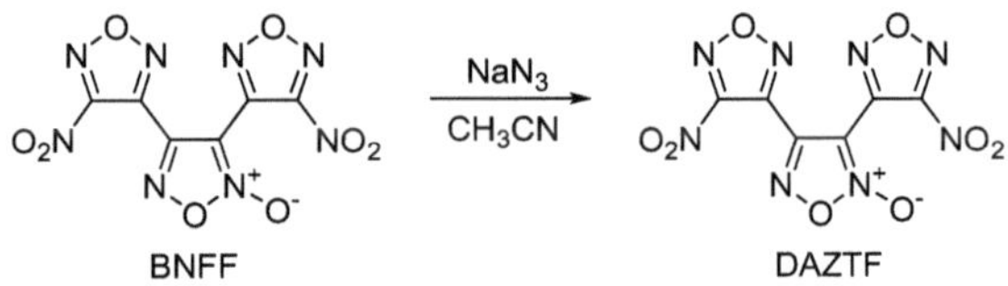

BNFF is a powerful explosive with a relatively low melting point (108–110°C) and a performance similar to that of HMX [19]. It is reported to have a density of 1.937 g/cm^3 and a detonation velocity of 9.25 km/s. Because of these properties, BNFF has been considered for melt-cast explosives applications [20]. It has also been shown to form eutectics with a variety of other energetic materials such as TNT, pentaerythritol tetranitrate, and trinitroazetidine.

The nitro groups on BNFF are labile and reactive toward nucleophiles such as amines and azides. Treatment of BNFF with sodium azide in acetonitrile results in the bisazido compound DAZTF (Scheme 16) [21]. DAZTF has a density of 1.743 g/cm^3, a melting point of 50–52°C, and a thermal decomposition onset temperature of 204°C. The detonation velocity was estimated to be 8.5 km/s.

5.3 Macrocyclic Furoxans

In 2012, Chavez and coworkers reported the synthesis of a macrocycle prepared from the oxidation of bis(aminofurazanyl)furoxan using trichloroisocyanuric acid as the oxidant (Scheme 17) [22]. The resulting product has the possibility of forming two isomers.

Mahkova and coworkers synthesized a similar set of furoxan macrocycles [23]. The starting furoxan substrates are based on *N*-oxide isomers of 3-amino-4-(aminofurazanyl)furoxan. The position of the *N*-oxide was either internal or external to the heterocycle–heterocycle bond. Scheme 18 displays the resulting heterocycles formed after oxidation. In the case of the external *N*-oxide, only one macrocycle was isolated, when the diamine was oxidized with dibromoisocyanuric acid (DBI). With the internal *N*-oxide, oxidation with DBI resulted in

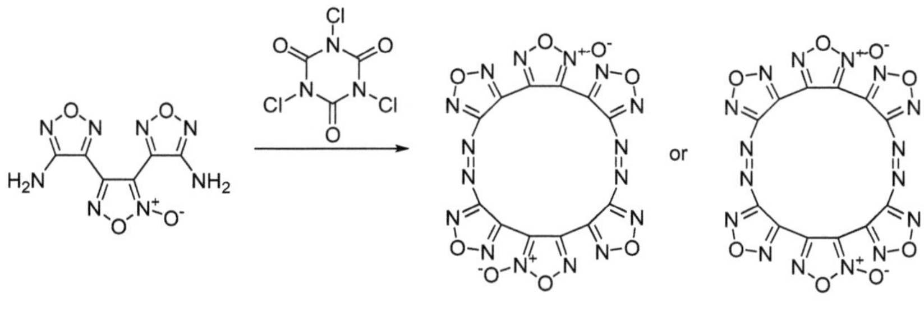

Scheme 17 Bis-furazanyl-furoxan macrocycle synthesis

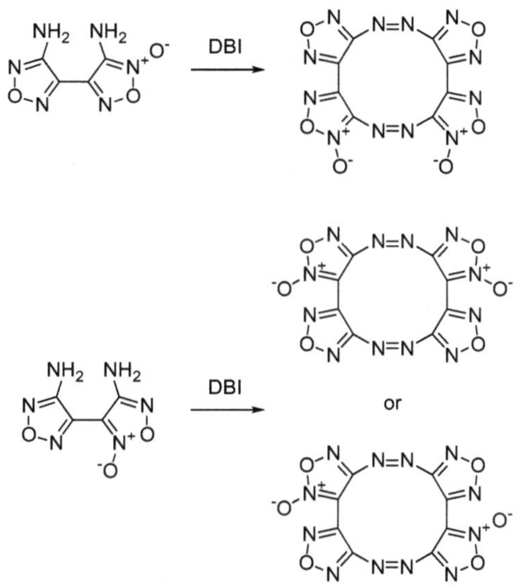

Scheme 18 Furazanyl-furoxan macrocycle synthesis

a 12-membered macrocycle, but the assignment of which isomer was formed was not determined. These macrocycles were formed in low yield due to numerous other oligomeric products that are formed in the process.

Treatment of bis(nitrofurazanyl)furoxan (BNFF) with sodium carbonate in acetonitrile leads to cyclization to the bifurazano[3,4-*b*:3′,4′-*f*]furoxano[3″,4″-*d*] oxacycloheptatriene (BFFO) (Scheme 19) [24]. The product melts at 97°C and thus has potential as a melt-cast material. X-ray crystallographic analysis showed that the crystal density of the BFFO monohydrate was 1.866 g/cm^3. Solvate-free crystals could not be obtained but were predicted to have a density of 1.9 g/cm^3 and a heat of formation of 275.2 kJ/mol. BFFO was predicted to have a detonation velocity of 8.6 km/s and a detonation pressure of 34.6 GPa. The material was also described as less sensitive to impact than HMX.

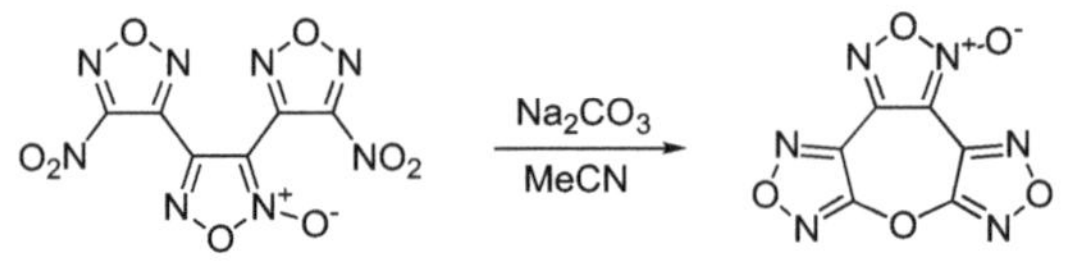

Scheme 19 Synthesis of BFFO

Scheme 20 Synthesis of azepines from BNFF

Treatment of bis(nitrofurazanyl)furoxan (BNFF) with ammonia in acetonitrile leads to cyclization to the bifurazano[3,4-*b*:3′,4′-*f*]furoxano[3″,4″-*d*]azacycloheptatriene (BFFO), while treatment with hydrazine provides the *N*-amino derivative (Scheme 20) [25]. The azepine compound is hygroscopic and forms a monohydrate crystal with a density of 1.817 g/cm^3 and a measured detonation velocity of 7.9 km/s. The *N*-amino compound is much more sensitive to impact and friction than the N–H azepine, which is thought to be due to the relative weakness of the N–N bond [26].

5.4 Other Furoxan Derivatives

The two isomers of 3,3′-diamino-4,4′-bifuroxan can be converted to the bisnitramine compounds by treatment with 100% nitric acid (Scheme 21) [27]. While these derivatives display poor thermal stability ($T_d < 95$°C), they could be converted to nitrogenous salts by treatment with ammonia, hydroxylamine, or hydrazine. The bisnitramine with the *N*-oxides external to bisfuroxan bond was also converted to the guanidine, diaminoguanidine, and triaminoguanidine salts. All of the materials prepared were characterized with respect to their energetic materials safety and performance properties. None of the salts displayed thermal stability above 168°C. The most promising salts were the hydroxylammonium compounds, which displayed promising safety properties and detonation performance properties ($V_D = 9.5$–9.8 km/s, $P_{CJ} = 41.5$–46.4 GPa).

In 2005, Hiskey and coworkers reported the synthesis of a fused, tricyclic furoxan, 7-nitrotetrazolo[1,5-*f*]furazano[4,5-*b*]pyridine 1-oxide [28]. The synthesis began with the nitration of 2,6-dimethoxy pyridine, which gave 3,5-dinitro derivative, followed by subsequent treatment with hydrazine and diazotization to give 2,6-diazido-3,5-dinitropyridine (Scheme 22). Azidoazine compounds are known to participate in azido-tetrazolo tautomerization. In the case of 2,6-diazido-3,5-

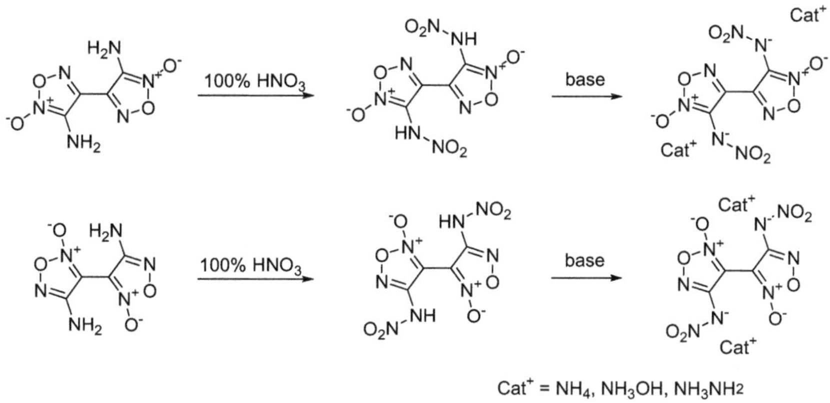

Scheme 21 Synthesis of dinitraminebisfuroxans and corresponding nitrogenous salts

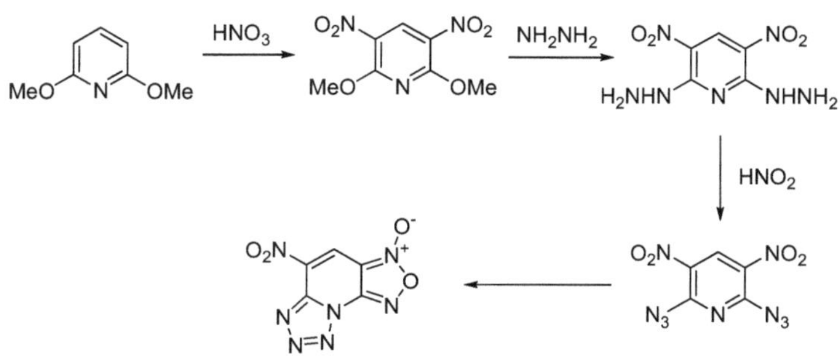

Scheme 22 Synthesis of 7-nitrotetrazolo[1,5-*f*]furazano[4,5-*b*]pyridine 1-oxide

dinitropyridine, dissolution in deuterated acetonitrile resulted in tautomerization to the tetrazolo form, followed by loss of dinitrogen and the formation of the furoxan ring. The diazido compound undergoes the transformation more quickly in polar solvents and converts to the product in the solid state over 8–10 days. X-ray crystallography showed that the product has at least two polymorphs (alpha (density = 1.853 g/cm^3), beta (density = 1.828 g/cm^3)). The material began to decompose at 160°C, was insensitive to electrostatic discharge, displayed moderate sensitivity to friction, and was slightly more sensitive to impact than HMX (octahydro-1,3,5,7-tetranitro-1,3,5,7-tetrazocine).

5.5 3-Oxyfuroxan

3-Oxyfuroxan is predicted to have higher heat of formation than furoxan while also having improved oxygen balance [29]. Thus, energetic derivatives of 3-furoxan may show promise as high-performing materials. Chen and coworkers recently published the preparation of energetic metal organic frameworks incorporating a 3-oxyfuroxan. In the work, a bis(oximylchloride)furoxan was treated with dinitrogen pentoxide, and this resulted in a mixture of the bis(dinitrochloromethyl)furoxan and bis(dinitrochloromethyl)-3-oxyfuroxan (Scheme 23). Treatment of this mixture with KI in methanol followed by reaction with silver nitrate produced a metal organic framework in which the 3-oxyfuroxan derivative can be identified. No properties of the pure 3-oxyfuroxan compound were reported.

6 Energetic Pyridine *N*-Oxides

6.1 Aminonitropyridine N-*Oxides*

Energetic pyridine *N*-oxides have been reported by Wilson [30]. 2,6-Diamino-3,5-dinitropyridine can be oxidized with 30% hydrogen peroxide in acetic acid (Scheme 24). The reaction proceeds to give an 84% yield of the *N*-oxide product. This material is very thermally stable and does not begin to decompose until 340°C. X-ray crystallography experiments showed that the density of the *N*-oxide is 1.878 g/cm^3. Treatment of the *N*-oxide with hydroxylamine in the presence of potassium hydroxide gives 2,4,6-triamino-3,5-dinitropyridine *N*-oxide. The density of the material remains nearly the same (1.876 g/cm^3) and yet is slightly less

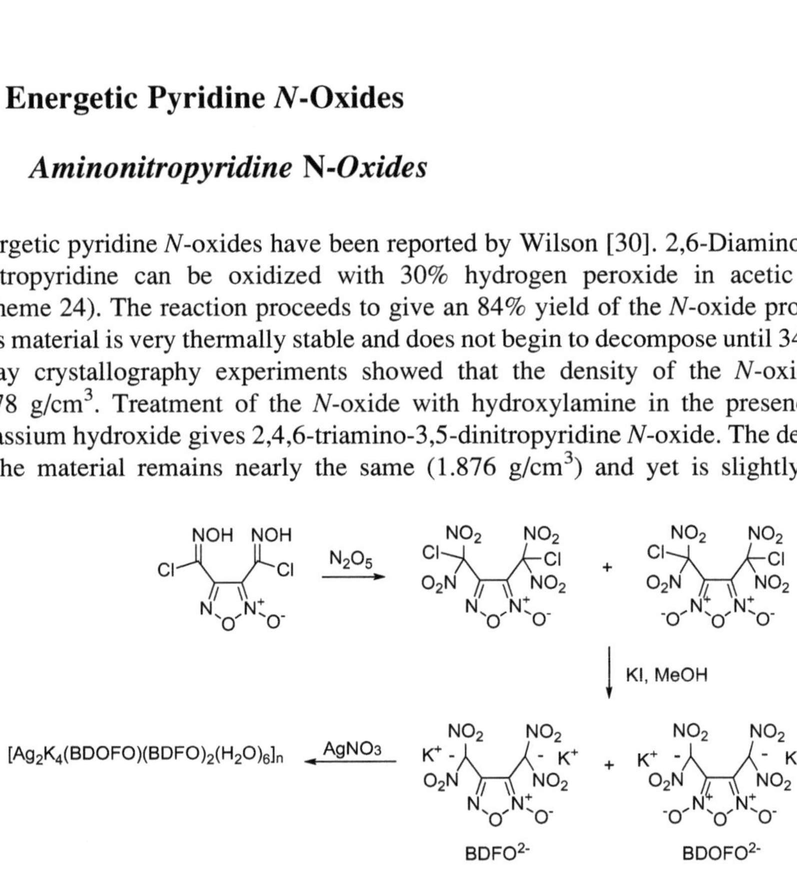

Scheme 23 Synthesis of a metal organic framework incorporating a 3-oxyfuroxan derivative

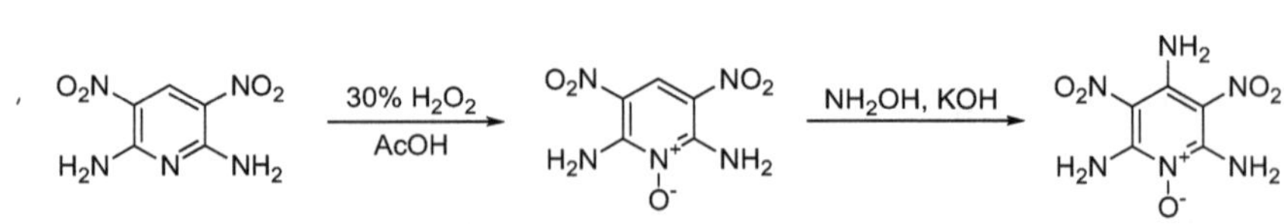

Scheme 24 Synthesis of dinitramine bisfuroxans and corresponding nitrogenous salts

thermally stable (308°C). The authors of this study also attempted to prepare the 2,6-dinitro-3,5-diaminopyridine 1-oxide but were unsuccessful.

7 Pyrazine *N*-Oxides

7.1 2,6-Diamino-3,5-Dinitropyrazine 1-Oxide

In the 1990s, scientists at the Lawrence Livermore National Laboratory described the synthesis of 2,6-diamino-3,5-dinitropyrazine 1-oxide (LLM-105) [31]. LLM-105 displays a crystal density of 1.919 g/cm^3 [32] and is described as an insensitive energetic material, with a detonation velocity of 8.73 km/s and a detonation pressure of 35.9 GPa [33].

The synthesis route is displayed in Scheme 25. The synthesis began with the nitration of 2,6-dimethoxypyrazine using mixed acid nitration conditions. The 2,6-dimethoxy-3,5-dinitropyrazine is then treated with ammonium hydroxide to provide 2,6-diamino-3,5-dinitropyrazine (ANPZ). Conversion to the *N*-oxide is performed using peroxytrifluoroacetic acid (PTFA). Unfortunately, the ANPZ and LLM-105 have low solubility in most solvents, and as the oxidation progresses, starting material becomes occluded in the *N*-oxide product. The overall result is a product that is 92–94% LLM-105, with the remained being unoxidized ANPZ. The ANPZ impurity does lead to a slightly lower explosive performance.

In 2010, an alternative route to LLM-105 was reported in a patent application [34]. The new route involves the treatment of bis(cyanomethyl)nitrosamine with hydroxylamine (Scheme 26), resulting in a cyclization process that provides 2,6-diaminopyrazine 1-oxide (DAPO) as the product. The material is then nitrated using mixed acid to provide LLM-105. The DAPO route now allows for LLM-105 to be free from the ANPZ impurity.

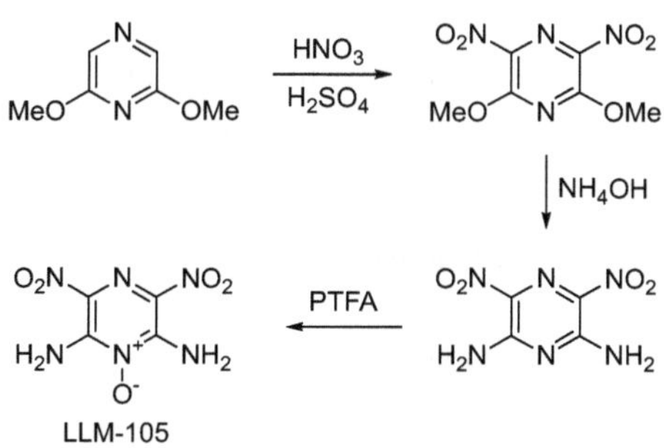

Scheme 25 Synthesis of LLM-105

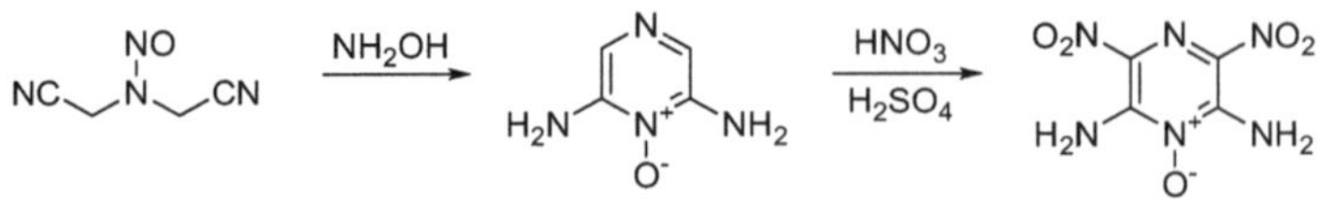

Scheme 26 DAPO route to LLM-105

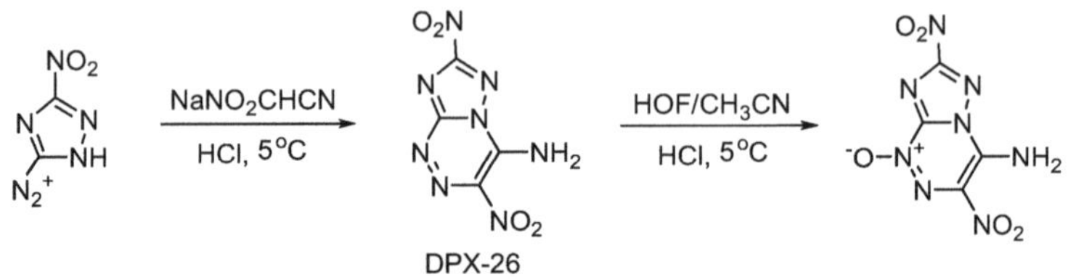

Scheme 27 The synthesis of 4-amino-3,7-dinitrotriazolo[5,1-*c*] [1,2,4]triazine 4-oxide

8 1,2,4-Triazine Oxides

*8.1 4-Amino-3,7-Dinitrotriazolo[5,1-*c*][1,2,4]Triazine 4-Oxide*

Examples of energetic 1,2,4-triazine *N*-oxides are rare. Recently, Piercey reported the synthesis of an energetic triazolo-1,2,4-triazine 4-oxide [35]. The synthesis involves a very interesting cyclization reaction that proceeds after condensation of a diazonium salt with nitroacetonitrile, resulting in the 4-amino-3,7-dinitrotriazolo[5,1-*c*] [1,2,4]triazine (Scheme 27). Oxidation of this substrate proved challenging, as strong oxidants such as peroxytrifluoroacetic acid and Oxone were unsuccessful. Ultimately, the use of hypofluorous acid was effective at installing an *N*-oxide at the 4-position of the bicyclic ring system. The material was predicted to have a density of 1.904 g/cm^3 and a heat of formation of 378 kJ/mol. The crystal structure of the dihydrate or nitromethane solvate shows that the triazine ring bonds to the triazole have lengthened in the *N*-oxide compound, and this may explain the large drop in thermal stability (T_{dec} = 138 vs 232°C). This triazine *N*-oxide was predicted to be a high-performing material that is less sensitive to impact and friction than conventional high-performance explosives (V_D = 8.97 km/s, P_{CJ} = 35.4 GPa).

9 1,2,3,4-Tetrazine *N*-Oxides

9.1 1,2,3,4-Tetrazine 1-Oxide

The synthesis of 1,2,3,4-tetrazine 1-oxide was reported by Piercey and coworkers in 2015 [36]. The synthesis involves the nitrene insertion of the *N*-amino-1,2,3-

triazole compound shown in Scheme 28. Treatment of this compound with MnO_2 resulted in the 1,2,3,4-tetrazine-1-benzyloxy *p*-toluenesulfonate salt. Subsequent treatment with hydrogen gas in the presence of Pd/C produced 1,2,3,4-tetrazine 1-oxide. The compound was confirmed by ^{13}C NMR and by mass spectrometry (DEI^+).

9.2 *1,2,3,4-Tetrazine Di-N-Oxides*

A review of work regarding annulated 1,2,3,4-tetrazine di-*N*-oxides has been published previously [37]. Since then, some very interesting work on annulated and non-annulated 1,2,3,4-tetrazine di-*N*-oxides has been published. In 2015, Churakov reported the thermolysis of poly azido benzotetrazine 1,3-dioxide in toluene. This results in the non-annulated dicyano-1,2,3,4-tetrazine 1,3-dioxide as the product (Scheme 29) [38]. Similarly, when 6,7- or 6,8-dimethoxybenzotetrazine 1,3-dioxides were treated with ozone, non-annulated 1,2,3,4-tetrazine 1,3-dioxides were produced (Scheme 30).

In 2014 Churakov reported the synthesis of 1,2,3,4-tetrazine 1,3-dioxides annulated with 1,2,3-triazoles and 1,2,3-triazole 1-oxides [39]. In this work, the tetrazine 1,3-dioxides are installed by condensing a *tert*-butylazoxy functional group with an amino group in the presence of sulfuric acid, nitric acid, and acetic anhydride (Scheme 31).

To access the 1,2,3-triazole 1-oxide tetrazine di-*N*-oxide derivatives, the same reaction conditions were employed using 1,2,3-triazole 1-oxide substrates (Scheme 32).

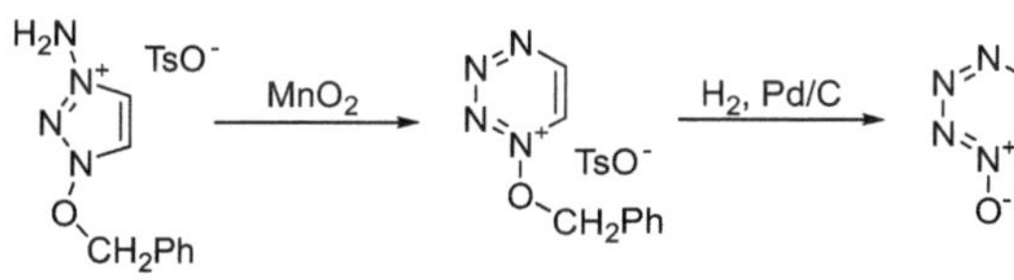

Scheme 28 The synthesis of 1,2,3,4-tetrazine 1-oxide

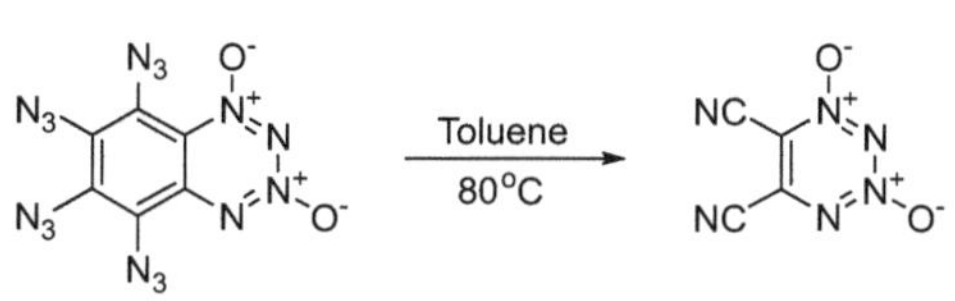

Scheme 29 Thermolysis of polyazidobenzotetrazine 1,3-dioxide

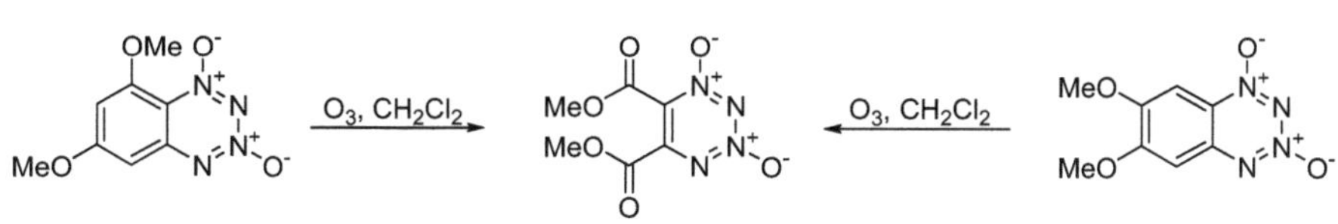

Scheme 30 Ozonolysis of dimethoxybenzotetrazine 1,3-dioxide

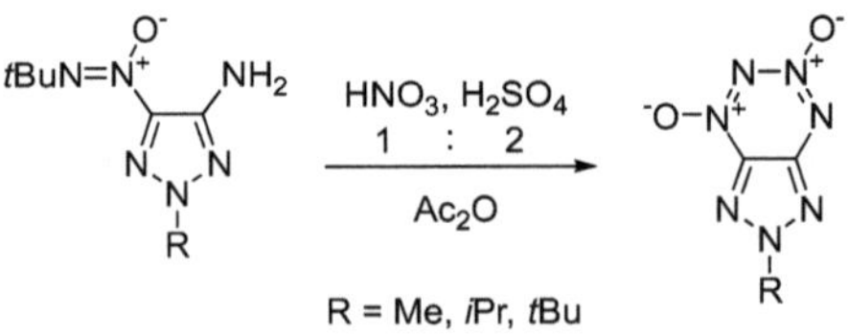

Scheme 31 Synthesis of 1,2,3-triazoletetrazine 1,3-dioxide

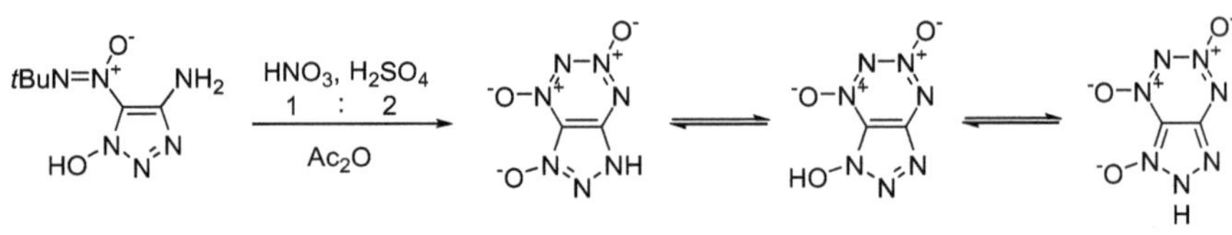

Scheme 32 Synthesis of 1-hydroxy-1*H*-[1,2,3]triazolo[4,5-*e*] [1,2,3,4]tetrazine 5,7-dioxide

Interestingly, non-alkylated 1,2,3-traizoles could also be used as substrates. The product, with an acidic proton, was predicted to be most stable in the 1-oxide 3-NH form. The materials were reported to display thermal stability between 154 and 199°C for the 1,2,3-triazoles derivatives and 180–230°C for the 1,2,3-triazole 1-oxide derivatives. No energetic materials properties for these derivatives were reported.

9.3 *Tetrazinotetrazine 1,3,6,8-Tetraoxide*

Two isomers of the fused bicyclic heterocycle, tetrazinotetrazine tetraoxide (TTTO), have been the targets of energetic materials synthesis chemist for decades (Fig. 1). The interest in these materials stems from the potential for these materials to display high densities and promising energetic materials properties. Additionally, these materials belong to a class known as alternating charge compounds. These materials have been predicted to display enhanced stability [40]. TTTO has been predicted to have a heat of formation of 206 kcal/mol and a predicted density of 1.98 g/cm^3. *Iso*-TTTO was also predicted to display similar properties [41]. Given these predicted properties, both materials are predicted to have a detonation velocity around 9.7 km/s and a detonation pressure of 432 kbar.

In 2016, Churakov and coworkers reported the first synthesis of TTTO [41]. They used a sequential ring closure strategy to install both ring systems by employing the process of annulating neighboring amino and (*tert*-butyl-*NNO*-azoxy) groups. The reported synthesis starts with bis (*tert*-butyl-*NNO*-azoxy)-cyano methane, which is converted to the cyclization precursor in seven steps (Scheme 33). The precursor undergoes the first annulation through treatment with $BF_3{\cdot}Et_2O$. Subsequent treatment with ammonia, followed by treatment with mixed acid in the presence of acetic anhydride, produces TTTO.

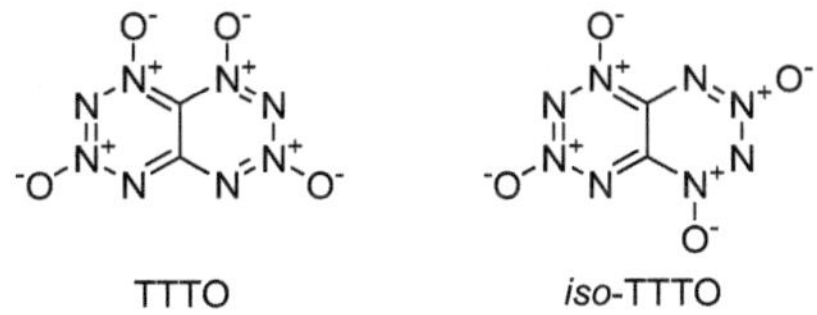

Fig. 1 Isomers of tetrazinotetrazine oxides

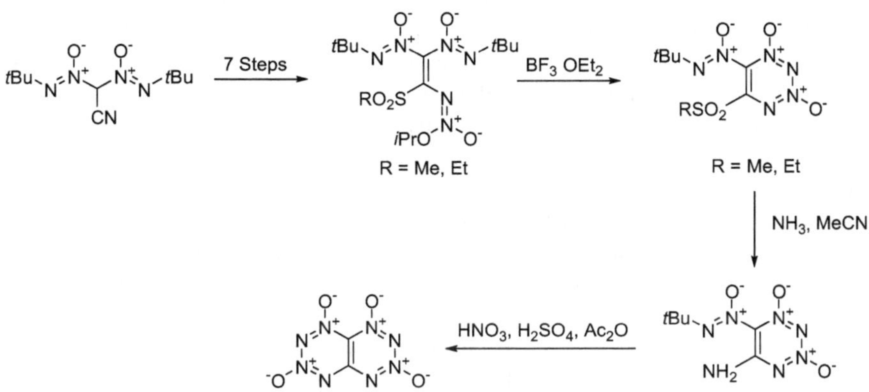

Scheme 33 The synthesis of TTTO

Unfortunately, TTTO was isolated as the minor product in the last step (23%). TTTO was obtained as a yellow powder with a melting point of 183–186°C. The structure was proven using X-ray crystallographic analysis of the benzene solvate of TTTO. Interestingly, the hydrolytic stability of TTTO and its benzene solvate is poor. While it is stable in air for a short time, it hydrolyzes in 50% aqueous ethanol after 2 h.

10 1,2,4,5-Tetrazine *N*-Oxides

10.1 3,6-Diaminotetrazine 1,4-Dioxide

One of the earliest heterocyclic *N*-oxides considered for energetic materials applications was 3,6-diamino-1,2,4,5-tetrazine 1,4-dioxide, often referred to as LAX-112 [42]. LAX-112 was synthesized by the oxidation of 3,6-diamino-1,2,4,5-tetrazine using Oxone as the oxidant (Scheme 34). The material has a crystal density of 1.86 g/cm^3 and a heat of formation of 39 kcal/mol. It is thermally stable up to 206°C, is insensitive to friction, and is relatively insensitive toward impact (26.2 J). LAX-112 can be formulated with a binder such as Kel-F or Viton

A. A 95:5 mixture of LAX-112:Viton A produces a material with a measured detonation velocity of 8.1 km/s and a detonation pressure of 30.1 kbar (unpublished results).

10.2 3-Amino-6-Nitrotetrazine N-*Oxides*

If 3,6-diamino-1,2,4,5-tetrazine is oxidized with peroxytrifluoroacetic acid (PTFA) instead, the mono-*N*-oxide is formed, while LAX-112 was not observed. It is believed that the PTFA was a strong enough acid to protonate the tetrazine ring of the mono-*N*-oxide and inhibit further oxidation to LAX-112. This material has a density of 1.76 g/cm^3 and is less thermally stable than LAX-112 ($T_{dec} = 150°C$) (Scheme 35). The experimental detonation performance of the material was not determined, but it is predicted to have a detonation velocity of 8.92 km/s and a detonation pressure of 32.5 GPa. A minor product was also isolated from the PTFA oxidation, namely 3-amino-6-nitro-1,2,4,5-tetrazine 2,4-dioxide and was likely to have arisen from a different oxidation pathway, involving oxidation of the amino group as the first step. This material was reported to have a density of 1.919 g/cm^3 but began to decompose at a much lower temperature (110°C). This material was predicted to have a detonation velocity of 9.33 km/s and a detonation pressure of 40.2 GPa (unpublished results).

In 2008, Mahkova and Ovchinnikov prepared 3-amino-6-nitro-1,2,4,5-tetrazine 2,4-dioxide through an alternative synthesis route, by diazotizing 3,6-diamino-1,2,4,5-tetrazine (Scheme 36) [43]. The method involved nonstandard diazotization conditions (DMSO, $NaNO_2$, HCl) to provide 3-amino-6-nitro-1,2,4,5-tetrazine followed by treatment with PTFA to give the di-*N*-oxide product. If less than a stoichiometric amount of PTFA is used, a mixture of the 2,4-dioxide and 2-oxide is formed in a 9:1 ratio. The use of an excess of oxidant results in only the 2,4-dioxide product. Interestingly, this study found that 3-amino-6-nitro-1,2,4,5-tetrazine

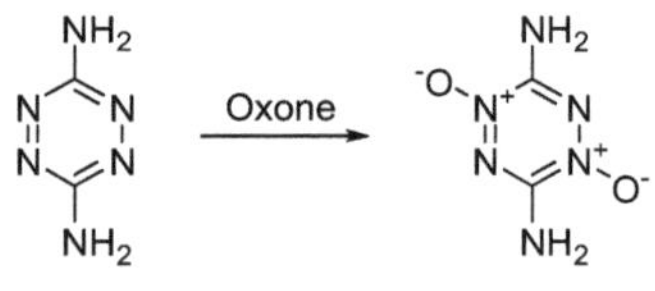

Scheme 34 Synthesis of LAX-112

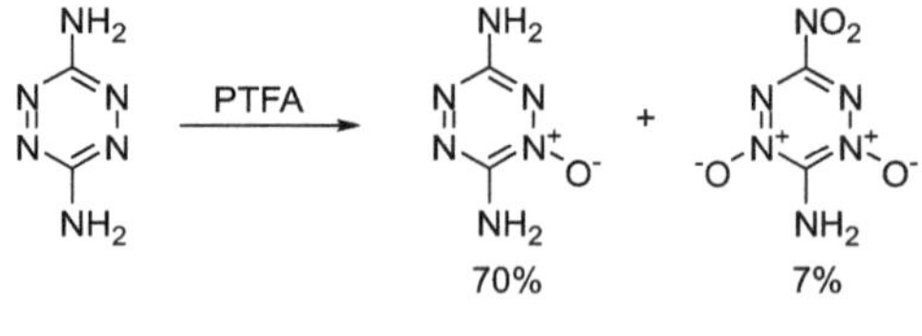

Scheme 35 Synthesis of 3,6-diamino-1,2,4,5-tetrazine 1-oxide and 3-amino-6-nitro-1,2,4,5-tetrazine 2,4-dioxide

2,4-dioxide was stable up to 191°C. No sensitivity or performance properties were reported for the mono-*N*-oxide compound.

In 1999, Hiskey and coworkers investigated hypofluorous acid as an oxidizer for 3,6-diamino-1,2,4,5-tetrazine [44]. Hypofluorous acid is one of the most powerful oxidizers and oxygen transfer reagents known [45]. The reaction proceeds to provide 3-amino-6-nitro-1,2,4,5-tetrazine 2,4-dioxide and 3-amino-6-nitro-1,2,4,5-tetrazine 1,4-dioxide, a new isomer (Scheme 37). This isomer was found to have a higher density (1.97 g/cm^3) and was found to be thermally stable up to 168°C. The predicted detonation velocity for this material is 9.81 km/s and the predicted detonation pressure is 42 GPa.

10.3 Other Amino-1,2,4,5-Tetrazine N*-Oxides*

The susceptibility of 1,2,4,5-tetrazine heterocycles toward oxidation strongly depends on the substituents at the 3 and 6 positions. If at least one of the tetrazine substituents is an amino group, the other substituent can be a hydrogen or a chlorine and oxidation can still occur with PTFA (Scheme 38) [44].

Oxidation of 3,6-bis(guanidinyl)-1,2,4,5-tetrazine with Caro's acid proceeds to provide the 1,4-dioxide product [46]. The guanidinyl functional groups of the product maintain enough basicity to form salts with strong acids (HNO_3 and $HClO_4$) (Scheme 39). The dinitrate salt was only stable to 157°C and was not studied further, but the perchlorate salt was stable up to 197°C. The perchlorate salt has a measured density of 1.94 g/cm^3 and a heat of formation estimated at −250 kJ/mol. The impact sensitivity was determined to be 3.7 J, and the friction sensitivity was

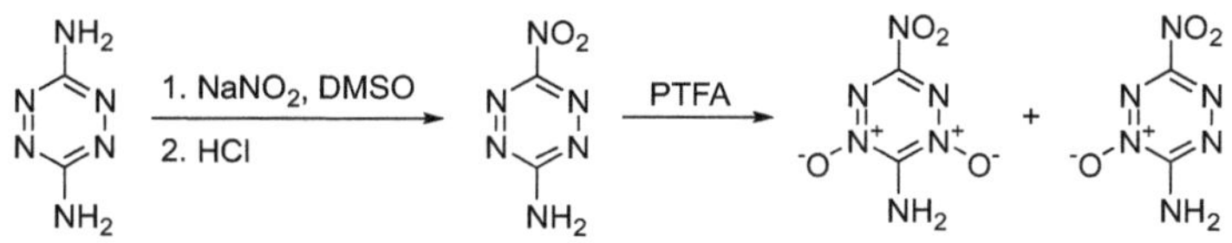

Scheme 36 Alternate synthesis of 3-amino-6-nitro-1,2,4,5-tetrazine 1-oxide and 3-amino-6-nitro-1,2,4,5-tetrazine 2,4-dioxide

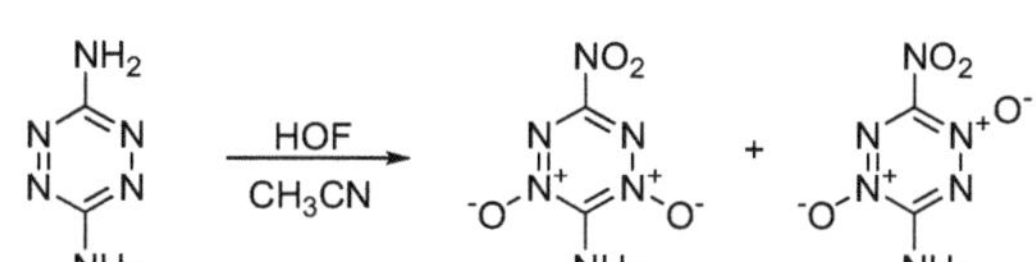

Scheme 37 Synthesis of 3-amino-6-nitro-1,2,4,5-tetrazine 2,4-dioxide and 3-amino-6-nitro-1,2,4,5-tetrazine 1,4-dioxide using hypofluorous acid

68 N. The detonation velocity of the perchlorate salt was experimentally determined to be 7.85 km/s.

3,6-bis((tetrazol-5-yl)amino)-1,2,4,5-tetrazine can also be oxidized with Caro's acid [46]. The structure was confirmed by X-ray crystallography to be the 1,4-dioxide (Scheme 40). Unfortunately, the thermal stability was poor (134°C) and further investigation was not performed.

In 2014, Shreeve reported the synthesis of 11 new 1,2,4,5-tetrazine mono- and di-*N*-oxides [47]. Figure 2 displays four of the more energetic compounds prepared in this work. The guanidine derivative has a density of 1.91 g/cm^3 and a calculated heat of formation of 325 kJ/mol. The explosive performance was predicted to be high ($V_D = 9.1$ km/s, $P_{CJ} = 37.5$ GPa), while the material was relatively insensitive (impact = 20 J, friction = 240 N).

10.4 *Oxides of 3,3′-Azobis(1,2,4,5-Tetrazine)*

The oxidation of 3,3′-azobis(6-amino-1,2,4,5-tetrazine) with excess PTFA results in a complex mixture of oxides [46]. Due to the insolubility of the product, characterization was limited to elemental analysis. Overall the product was found to contain 3.5 oxygens on average. The structures of those *N*-oxides are displayed in Scheme 41. The nitrogen atoms most likely to be converted to *N*-oxide are those adjacent to the amino groups as well as one of the two nitrogen atoms in the azo group. Unfortunately, the insolubility of the material precluded ^{15}N NMR analysis from being performed. The material displayed a gas pycnometry density of 1.88 g/cm^3 and a heat of formation of 150 kcal/mol. With these data, the predicted explosive performance is 9.0 km/s and 35.9 GPa. The combustion properties of

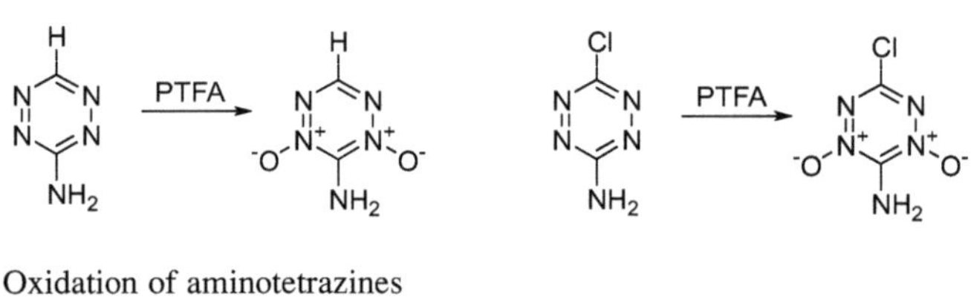

Scheme 38 Oxidation of aminotetrazines

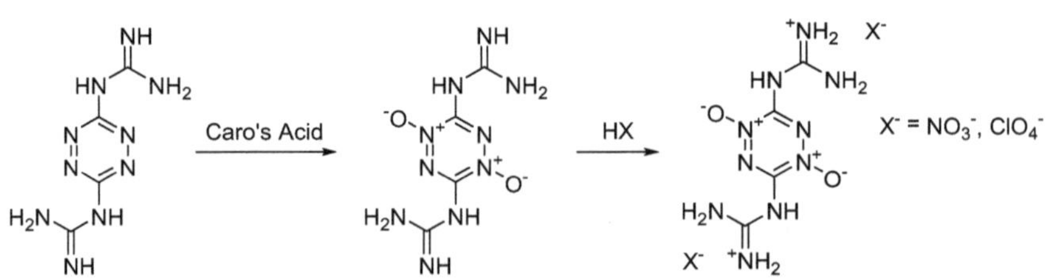

Scheme 39 Synthesis of 3,6-bis(guanidinyl)-1,2,4,5-tetrazine 1,4-dioxide and salts

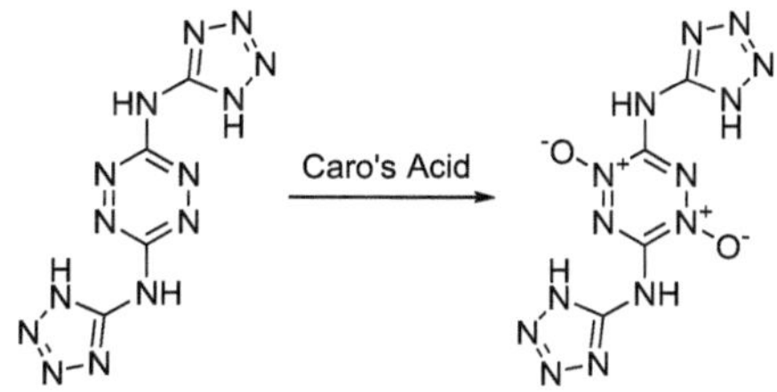

Scheme 40 Synthesis of 3,6-bis((1*H*-tetrazol-5-yl)amino)-1,2,4,5-tetrazine 1,4-dioxide

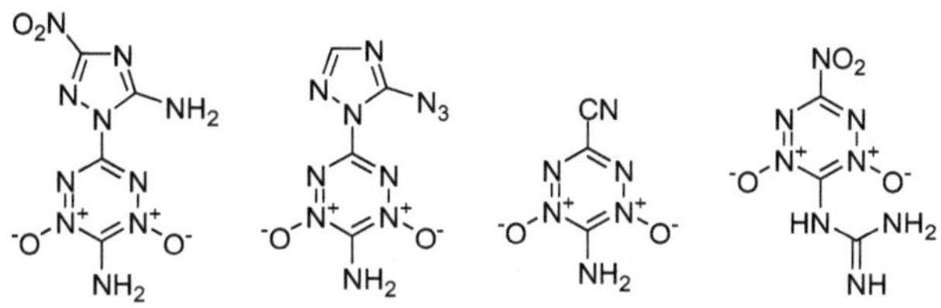

Fig. 2 Other amino-1,2,4,5-tetrazine 2,4-dioxides

Scheme 41 Reaction of 3,3′-azobis(6-amino-1,2,4,5-tetrazine) with PTFA

this material were determined. The material displayed exceptionally high burn rates and a low pressure dependence, two properties which are very important in propulsion applications. The burn rate was determined to be 6.6 cm/s at 14.4 MPa and the pressure dependence was found to be 0.27. At that time, it was the fastest burning material known in the literature. Because of these promising properties, the material has been studied extensively as a monopropellant in microthruster applications [48].

11 1,2,3,5-Tetrazine *N*-Oxides

11.1 Triazolo-1,2,3,5-Tetrazine 2-Oxide

1,2,3,5-Tetrazines are much less common than 1,2,4,5-tetrazines, and examples of energetic 1,2,3,5-tetrazine *N*-oxides are even more scarce. Zhou reported the synthesis of a novel energetic structure, 7-nitro-4-oxo-4,8-dihydro [1,2,4]triazolo [5,1-*d*] [1,2,3,5]tetrazine 2-oxide [49]. The compound was synthesized by treating 3-nitro-1-(2*H*-tetrazol-5-yl)-1*H*-1,2,4-triazol-5-amine with oleum and fuming nitric acid (Scheme 42). The proposed mechanism involves the formation

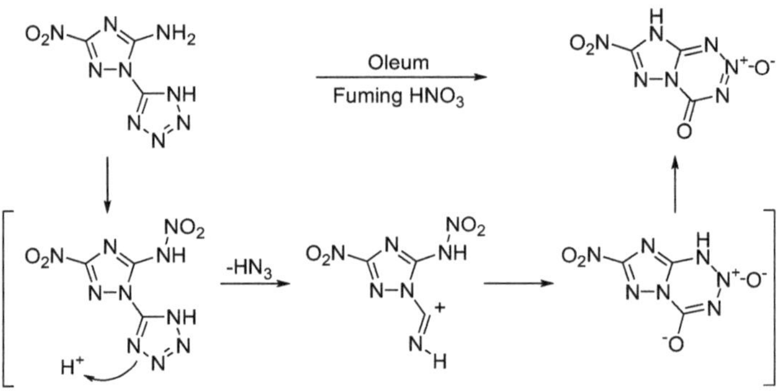

Scheme 42 Synthesis of 7-nitro-4-oxo-4,8-dihydro [1,2,4]triazolo[5,1-*d*] [1,2,3,5]tetrazine 2-oxide

of an intermediate nitramine, loss of hydrazoic acid, and subsequent condensation to form the product. The product is relatively acidic and can be converted to numerous nitrogen-based salts such as ammonium, hydroxylammonium, guanidinium, and others. The free acid was reported to be hygroscopic and its energetic and safety properties were not reported. All of the salts displayed thermal stability above 200°C, except for the hydroxylamine salt, which began to decompose at 197°C. The hydroxylamine salt also displayed the most promising properties, as it was reported to be insensitive to impact and friction, yet was predicted to be a high-performing compound ($V_D = 9.0$ km/s, $P_{CJ} = 39.5$ GPa). The high performance was in part due to the density (1.97 g/cm^3) measured using gas pycnometry.

References

1. Vinogradov VM, Dalinger IL, Ugrak BI, Shevelev SA (1996) Mendeleev Commun 6:139–140
2. Zhang Y, Parrish DA, Shreeve JMJ (2012) Mater Chem 22:12659–12665
3. Zelenov VP, Voronin AA, Churakov AM, Strelenko YA, Tartakovsky VA (2014) Russ Chem Bull Int Ed 63:123–129
4. Petrie MA, Koolpe G, Malhotra R, Penwell P (2012) Office of Naval Research Grant No.: N00014-08-0894. http://www.dtic.mil/get-tr-doc/pdf?AD=ADA561743
5. Starchenkov IB, Andrianov VG, Mishnev AF (1997) Chem Heterocycl Compd 33:1355–1359
6. Thottempudi V, Yin P, Zhang J, Parrish DA, Shreeve JM (2014) Chem A Eur J 20:542–548
7. Bottaro JC, Petrie MA, Penwell PE (2002) Integrated high payoff rocket propulsion technology III: advanced solution propellant technology, contract number: F04611-98-C-0018, final report
8. Dippold AA, Klapoetke TM (2013) J Am Chem Soc 135:9931–9938
9. Gobel M, Karaghiosoff K, Klapotke TM, Piercey DG, Stierstorfer JJ (2010) Am Chem Soc 132:17216–17226
10. Klapoetke TM, Piercey DG, Stierstorfer J (2011) Chem A Eur J 17:13068–13077
11. Fischer D, Klapoetke TM, Piercey DG, Stierstorfer J (2013) Chem A Eur J 19:4602–4613

12. Fischer N, Fischer D, Klapoetke TM, Piercey DG, Stierstorfer JJ (2012) Mater Chem 22:20418–20422
13. Tselinski IV, Mel'nikova SF, Romanova TV (2001) Russ J Org Chem 37:430–436
14. Klapoetke TM, Kurz MQ, Schmid PC, Stierstorfer J (2015) Energy Mater 33:201–215
15. Binnikov AN, Kulikov AS, Mahkova NN, Ochinnikov OV, Pivina TS (1999) 4-Amino-3-azidocarbonyl furoxan as a universal synthon for the synthesis of energetic compounds of the furoxan series. In: 30th international annual conference of ICT, Karlsrruhe, Germany, 58/1-58/10
16. Sheremetev AB, Ivanova EA, Spiridonova NP, Melnikova SF, Tselinsky IV, Suponitsky KY, Antipin MY (2005) J Heterocycl Chem 42:1237–1242
17. Zhao F-Q, Hen P, Hu R-Z, Luo Y, Zhang Z-Z, Zhou Y-S, Yang X-W, Gao Y, Gao S-L, Shi Q-Z (2004) J Hazard Mater A 113:67
18. Wang J, Li J, Liang Q, Huang Y, Dong H (2008) Propellants Explos Pyrotech 33:347–352
19. Stepanov AI, Dashko DV, Astrat'ev AA (2012) Cent Eur J Energetic Mater 9:329–342
20. Wang Q (2003) Chin J Explos Propellants 3:57–59
21. Zhou Y, Zhang Z, Li J, Guan X, Huang X, Zhou C (2010) Chin J Org Chem 30:1044–1050
22. Chavez DE, Parrish DA, Leonard P (2012) Synlett 23:2126–2128
23. Epishina MA, Kulikov AS, Mahkova NN (2008) Russ Chem Bull Int Ed 57:644–651
24. Zhou Y, Xu K, Wang B, Zhang H, Qui Q, Zhao F (2012) Bull Kor Chem Soc 33:3317–3320
25. Astrat'ev AA, Dashko DV, Stepanov AI (2012) Cent Eur J Energetic Mater 10:1087–1094
26. Sheremetev AB, Yudin IL (2003) Russ Chem Bull 72:87–100
27. He C, Tang Y, Mitchell LA, Parrish DA, Shreeve JMJ (2016) Mater Chem A 4:8969–8973
28. Huynh MHV, Hiskey MA, Chavez DE, Gilardi RDJ (2005) Energy Mater 23:99–106
29. Zhai L, Qu X, Wang B, Bi F, Chen S, Fan X, Xie G, Qing W, Gao S (2016) ChemPlusChem. doi:10.1002/cplu.201600287
30. Hollins RA, Merwin LH, Nissan RA, Wilson WSJ (1996) Heterocycl Chem 33:895–904
31. Pagoria PF, Mitchell AR, Schmidt RD, Simpson RL, Garcia F, Forbes J, Cutting J, Lee R, Swansiger R, Hoffman DM (1998) Synthesis scale-up and experimental testing of LLM-105 (2,6-diamino-3,5-dinitro-pyrazine-1-oxide). National Defense Industrial Association, San Diego
32. Pagoria PF (2016) Propellants Explos Pyrotech 41:452–469
33. Sabatini JJ, Oyler KD (2016) Crystals 6:5. doi:10.3390/cryst6010005
34. Pagoria PF, Zhang MX. US20100267955
35. Piercey DG, Chavez DE, Scott BL, Imler GH, Parrish DA (2016) Angew Chem Int Ed 55: 15315–15318
36. Piercey DG, Chavez DE, Heimsch S, Kirst C, Klapoetke TM, Stierstorfer J (2015) Propellants Explos Pyrotech 40:491–497
37. Churakov AM, Tartakovsky VA (2004) Chem Rev 104:2601–2616
38. Klenov MS, Churakov AM, Strelenko YA, Ananyev IV, Lyssenko KA, Tarakovsky VA (2015) Tetrahedron Lett 56:5437–5440
39. Voronin AA, Zelenov VP, Churakov AM, Strelenko YA, Feyanin IV, Tartakovsky VA (2014) Tetrahedron 70:3018–3022
40. Christe KO, Dixon DA, Vasilliu M, Wagner RI, Haiges R, Boatz JA, Ammon HL (2015) Propellants Explos Pyrotech 40:463–468
41. Klenov MS, Guskov AA, Anikin OV, Churakov AM, Strelenko YA, Fedyanin IV, Lyssenko KA, Tartakovsky VA (2016) Angew Chem Int Ed 55:11472–11475
42. Coburn MD, Buntain GA, Harris BW, Hiskey MA, Lee K-Y, Ott DG (1991) J Heterocycl Chem 28:2049
43. Ovchinnikov IV, Mahkova NN (2008) In: New trends in research of energetic material 2008, Pardubice, Czech Republic, pp 713–718
44. Chavez DE, Hiskey MAJ (1999) Energy Mater 17:357–377
45. Rozen S, Brand M (1986) Angew Chem Int Ed 25:554
46. Chavez DE, Naud DL, Hiskey MA (2004) Propellants Explos Pyrotech 29:209–215
47. Wei H, Gao H, Shreeve JM (2014) Chem A Eur J 20:16943–16952
48. Ali AN, Son SF, Hiskey MA, Naud DL (2004) J Propuls Power 20:120
49. Bian C, Dong X, Zhang X, Zhou Z, Li CJ (2015) Mater Chem A 3:3594–3601

Top Heterocycl Chem (2017) 53: 29–58
DOI: 10.1007/7081_2017 3

Published online: 19 April 2017

Pyridine *N*-Oxides and Derivatives Thereof in Organocatalysis

Petr Koukal, Jan Ulč, David Nečas, and Martin Kotora

Abstract Pyridine *N*-oxides are a class of mild Lewis bases that can activate certain kinds of Lewis acidic parts of molecules and hence increase reactivity of its nucleophilic part towards various reactions with electrophiles. This review aims to summarize the applications of various non-chiral and chiral pyridine *N*-oxides in catalysis of a variety of reactions. In addition, the applications of these reactions in syntheses of natural and bioactive compounds are mentioned as well.

Keywords Aldol reactions • Allylation • Catalysis • Crotylation • Cyanation • Epoxide • Epoxide cleavage • Natural compounds • *N*-oxide • Phosphorylation • Propargylation • Rearrangement • Reduction • Synthesis

Contents

P. Koukal, J. Ulč, D. Nečas, and M. Kotora (✉)
Department of Organic Chemistry, Faculty of Science, Charles University, Albertov 6, 128 43 Praha 2, Czech Republic
e-mail: martin.kotora@natur.cuni.cz

1 Introduction

Compounds possessing one, two, or more pyridine *N*-oxide moieties as the part of their molecular framework constitute a special class of compounds with unique properties. Probably the most frequent use of such compounds is in various oxidation reactions where they act as oxidants. However, this review aims to focus on the use of pyridine *N*-oxides as mild Lewis bases that can activate Lewis acid parts of molecules and hence increase reactivity of its nucleophilic part towards various reactions with electrophiles. Although catalytic activity of compounds possessing the pyridine *N*-oxide moiety has been partially reviewed or mentioned several times in different reports during the past two decades [1–13], the goal of this review is to gather a more comprehensive set of information on catalytic activity and applications of pyridine *N*-oxides in promotion of various racemic and enantioselective reactions.

First of all, it should be noted that pyridine *N*-oxides can be found in nature. Usually these compounds are highly toxic. As a typical example may serve orellanine **A** – a mycotoxin – isolated from the Fool's webcap (*Cortinarius orellanus*) (Fig. 1) [14, 15]. It is a highly poisonous substance with nephrotoxic properties. Poisoning results in renal failure and irreversible damage of kidneys. Other pyridine *N*-oxides possessing antibacterial activity such as 2-(methyldithio)pyridine-*N*-oxide **B**, 2-[(methylthiomethyl) dithio]pyridine-*N*-oxide **C**, and 2,2′-dithio-bis(pyridine-*N*-oxide) **D** were isolated from Persian shallot (*Allium stipitatum*) [16].

N-oxides are also found in the realm of pharmaceutically active substances (Fig. 2). As typical examples may serve SCH 350634 **E** and SCH 341125 **F** that serve as selective CCR5 receptor antagonists with potent anti-HIV activity [17], a potent thrombin inhibitor **G** [18, 19], and JPL-32 **H** a representative of *N*-oxides with anti-HIV properties with multiple mechanisms of antiviral action [20].

2 Physical Properties of Pyridine *N*-Oxides

Pyridine *N*-oxides are Lewis bases, because the N–O moiety of pyridine *N*-oxides, thanks to high polarization, might act as an electron donor. Hence they may combine with Lewis acids forming the corresponding Lewis acid–base pairs. This property has an essential chemical consequence, because it can increase the nucleophilicity of the Lewis acids towards potential electrophiles and thus allow them to react under conditions under which they otherwise would not react.

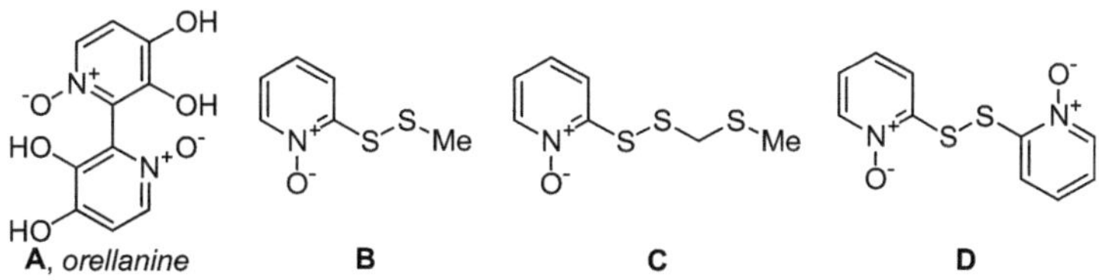

Fig. 1 Orellanine and other natural compounds possessing the *N*-oxide moiety

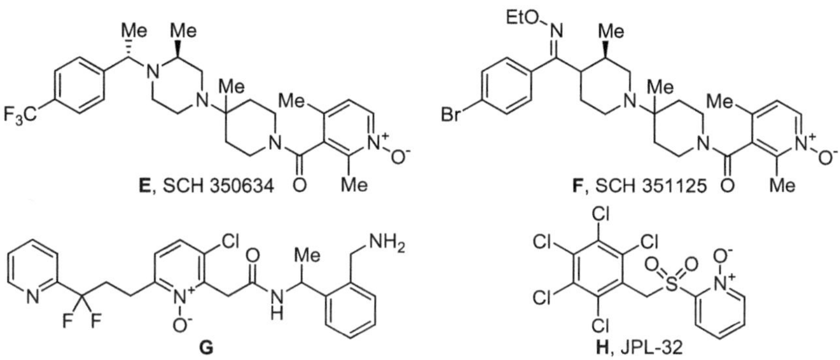

Fig. 2 Pharmaceutically active substances possessing the *N*-oxide moiety

Although the basicity of the pyridine *N*-oxides is lower in comparison with the corresponding pyridines, it is often sufficient enough to activate a number of various Lewis acids and thus catalyze a number of various reactions. Basicity of pyridine *N*-oxides and their respective reactivity towards numerous Brønsted and Lewis acids has been studied in a number of cases and for their pK_a and other data see elsewhere [21–26].

3 Non-chiral Pyridine *N*-Oxide Catalyzed Reactions

Only a handful of reactions catalyzed by non-chiral *N*-oxides have been reported. Among them belong allylation of aldehydes, aldol reaction, Passerini-type reaction, and phosphorylation of alcohols.

Polyisobutylene-supported pyridine *N*-oxide was shown to catalyze allylation of several aromatic aldehydes with allyltrichlorosilane (Scheme 1) [27]. These studies demonstrated that it is a highly active and recyclable catalyst that promotes the allylation of aromatic aldehydes in yields up to 99%. It could be successfully recycled up to five times by extraction with a mixture of hexane/90% EtOH–H_2O. The recycled catalyst retained its catalytic efficiency.

Pyridine *N*-oxide and DMAP *N*-oxide were used to catalyze aldol reaction of trimethylsilyl dimethylketene acetal with various aromatic (Table 1) and other aldehydes

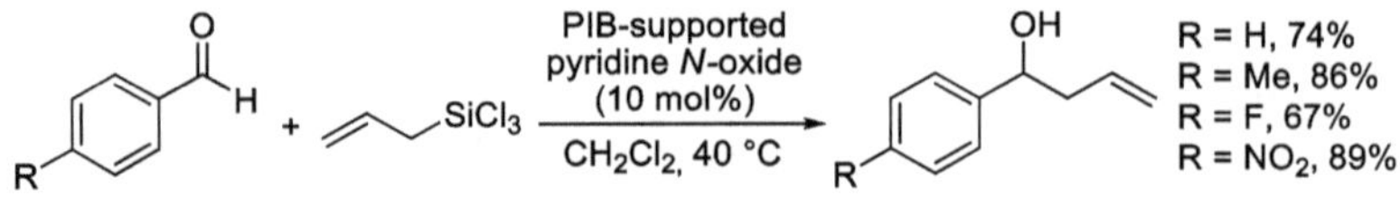

Scheme 1 Allylation of aldehydes catalyzed by a polymer supported pyridine *N*-oxide

Table 1 Aldol reactions with substituted benzaldehydes

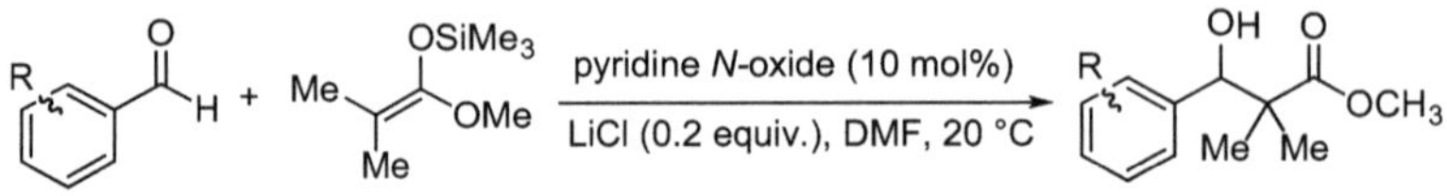

R	Time (h)	Yield (%)[a]
H	5	83
4-NO_2	23	87
2-Cl	12	65
4-Cl	10	77
4-MeO	10	44
4-HO	21	96
4-AcO	15	80
4-*t*-$BuMe_2SiO$	10	62
4-MeS	21	87

[a]Isolated yields

(Table 2) [28]. The reaction was carried out in DMF and in almost all cases it proceeded to furnish aldol products in high yields.

The Passerini reaction is based on the reaction of a mixture of an isocyanide with a ketone or an aldehyde in the presence of a carboxylic acid to give an α-acyloxy carboxamide. Pyridine *N*-oxide was shown to catalyze the reaction of $SiCl_4$ with benzaldehyde and *tert*-butyl isocyanide to the corresponding α-hydroxy amide after basic hydrolysis in 94% yield (Scheme 2) [29].

Recently, it has been shown that various 2-aryl-4-(dimethylamino)pyridine-*N*-oxides can serve as efficient phosphorylation catalysts for amino acids and polyols (Scheme 3) [30]. The most effective catalyst was 2-(2,4-bistrifluoromethylphenyl)-4-dimethylaminopyridine *N*-oxide. Synthetic usefulness of this procedure was also demonstrated by a selective phosphorylation of the tyrosine hydroxyl group in a heptapeptide.

As far as other reactions catalyzed by pyridine *N*-oxides are concerned, it is worth to mention: (1) a positive catalytic effect of 4-substituted pyridine *N*-oxides on silylation of alcohols [31, 32], (2) effect of pyridoxal *N*-oxide on racemization of amino acids [33], 4-substituted pyridine *N*-oxides for catalysis of hydrolysis of 2,3,5,6-*p*-benzoquinone [34], the use of pyridine *N*-oxide as a cocatalyst for arylsulfonylation and benzoylation of phenols [35, 36], and finally the use of 4-dimethylamino pyridine *N*-oxide as an efficient catalyst in peptide synthesis [37].

Table 2 Aldol reactions with other aldehydes

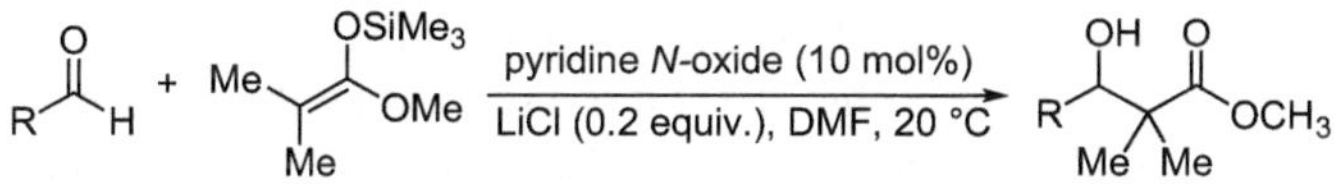

R	Time (h)	Yield (%)[a]
6-Methylpyrid-2-yl	5.5	55
$PhCH_2CH_2$	5	80
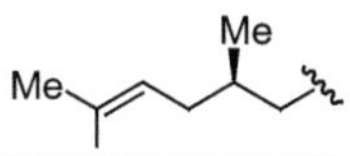	5	81
$Me(CH_2)_8$	7	78
4-*t*-$BuMe_2SiO$	10	62
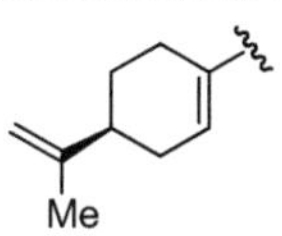	10	91

[a]Isolated yields

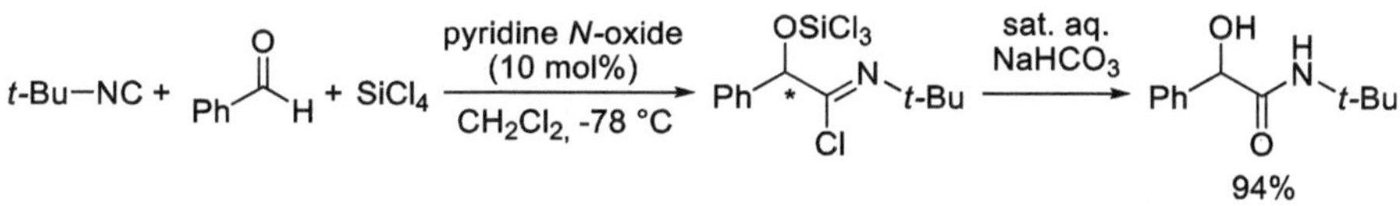

Scheme 2 Passerini-type reaction catalyzed by pyridine *N*-oxide

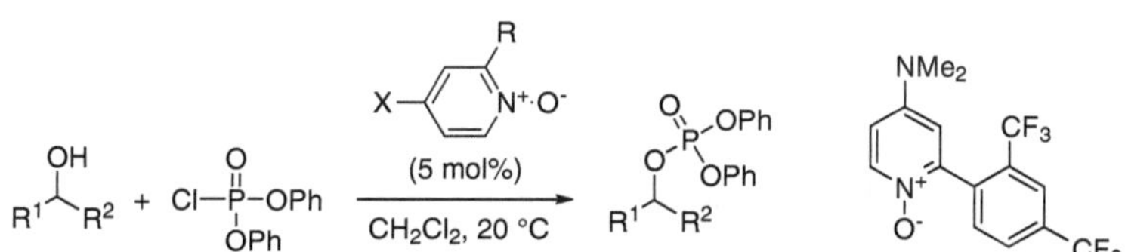

X = H, NMe_2
R = H, Cl, 4-$MeOC_6H_4$, 4-MeC_6H_4, 4-$MeOC_6H_4$, 3,5-$(CF_3)_2C_6H_3$, 2,4-$(CF_3)_2C_6H_3$

Scheme 3 Phosphorylation of alcohols catalyzed by a substituted pyridine *N*-oxide

4 Reactions Catalyzed by Chiral Pyridine *N*-Oxides and Their Derivatives

4.1 *Synthesis of Chiral Pyridine* N*-Oxides,* N,N′*-Dioxides, and Related Compounds*

Compounds possessing the pyridine *N*-oxide moiety could be divided into several classes with respect to their elements of chirality within their scaffold and can be classified into the following groups: (a) those possessing the element of axial chirality (Fig. 3), (b) those possessing the element of central chirality (Figs. 4–6), (c) those possessing the element of planar chirality (Fig. 7), (d) those possessing the element of helical chirality (Fig. 8), and, finally, (e) those possessing several elements of chirality (axial and central) (Fig. 9).

At the outset, the pyridine *N*-oxides possessing the element of axial chirality are dealt with. Historically the first chiral bipyridine *N*,*N*′-dioxide **1a** was prepared and resolved into enantiomers by Fujii et al. [38, 39] but it was Nakajima et al. [40] who

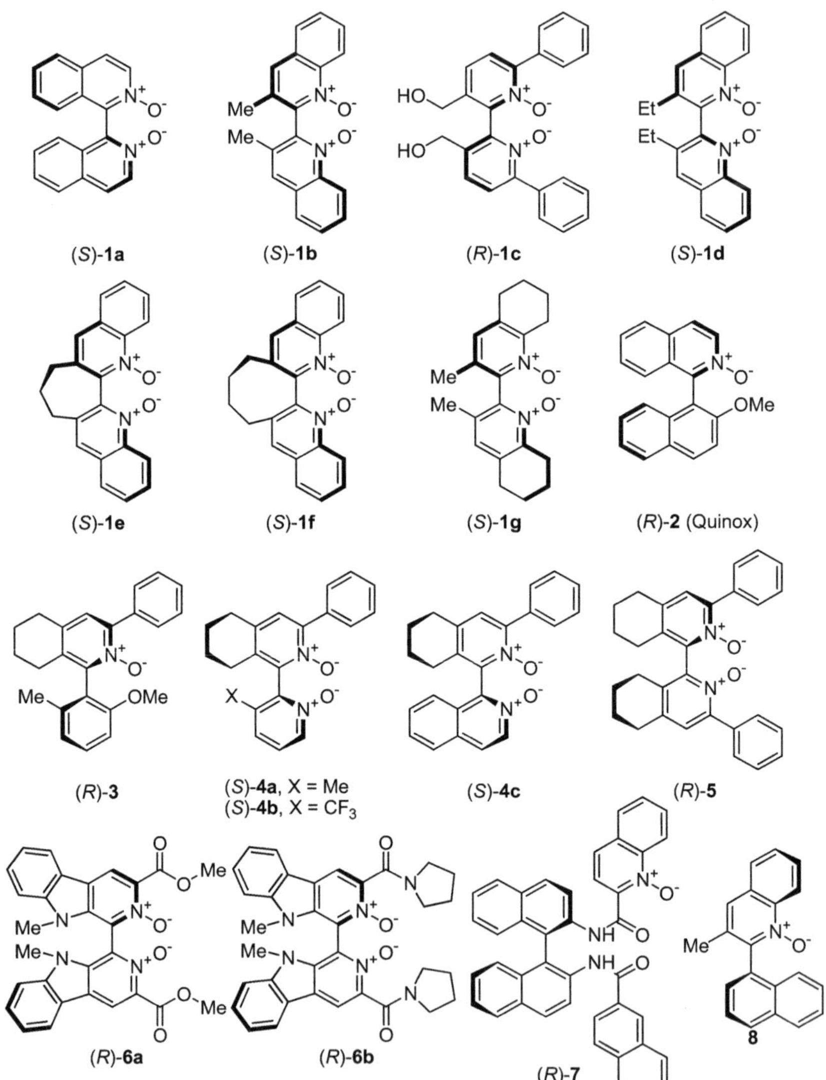

Fig. 3 Various pyridine *N*-oxides possessing elements of axial chirality

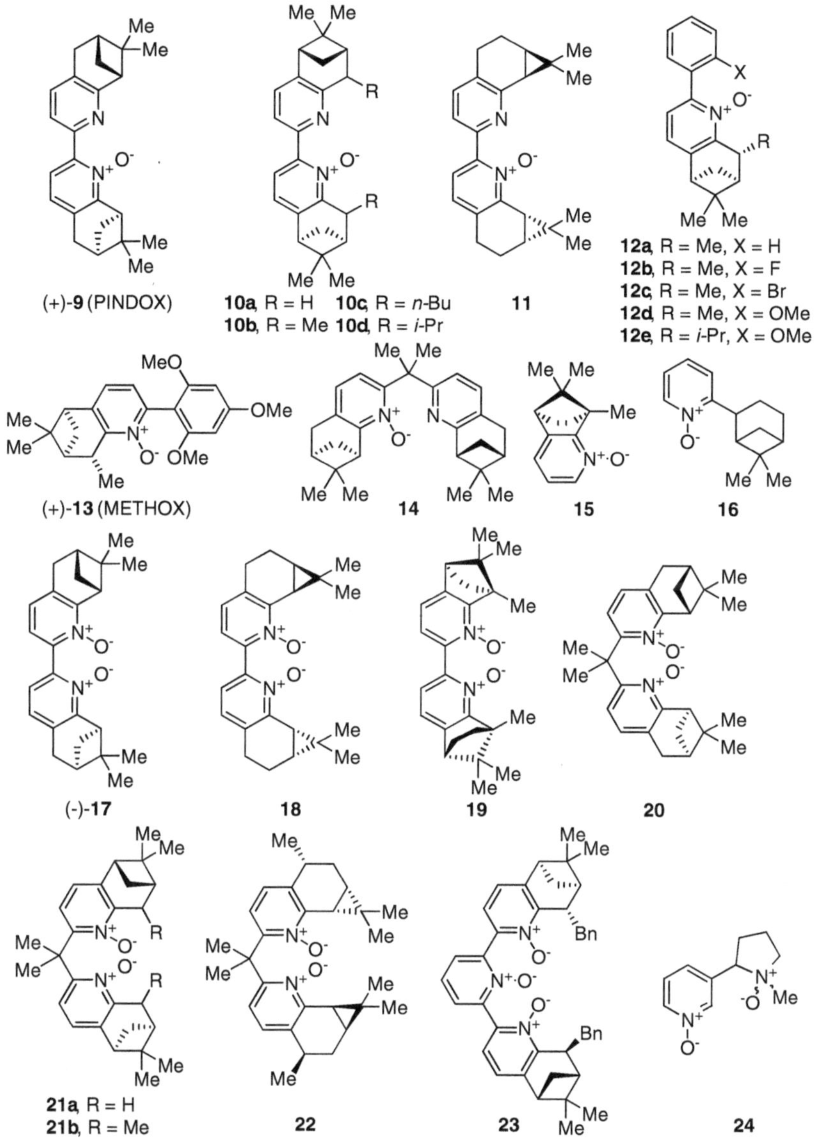

Fig. 4 Various pyridine *N*-oxides possessing elements of central chirality derived from terpenes and alkaloids

reported the first use of **1a** and its analog **1b** (Fig. 3) in organocatalysis. These achievements provided the necessary impetus that prompted others to join, expand, and develop the area of organocatalytic allylations into a matured field. The compound **1c** with a similar framework was prepared by Hayashi et al. [41, 42]. Other analogs such as **1d**, prepared by Chang et al. [43], and **1e–1g**, prepared by Feng et al. [44],

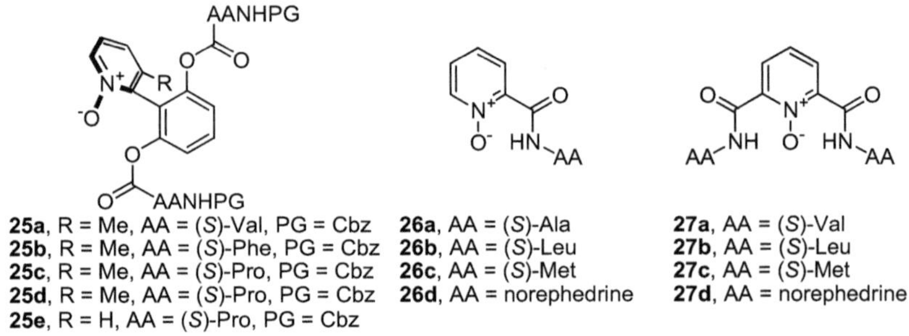

Fig. 5 Various pyridine *N*-oxides possessing elements of central chirality derived from amino acids

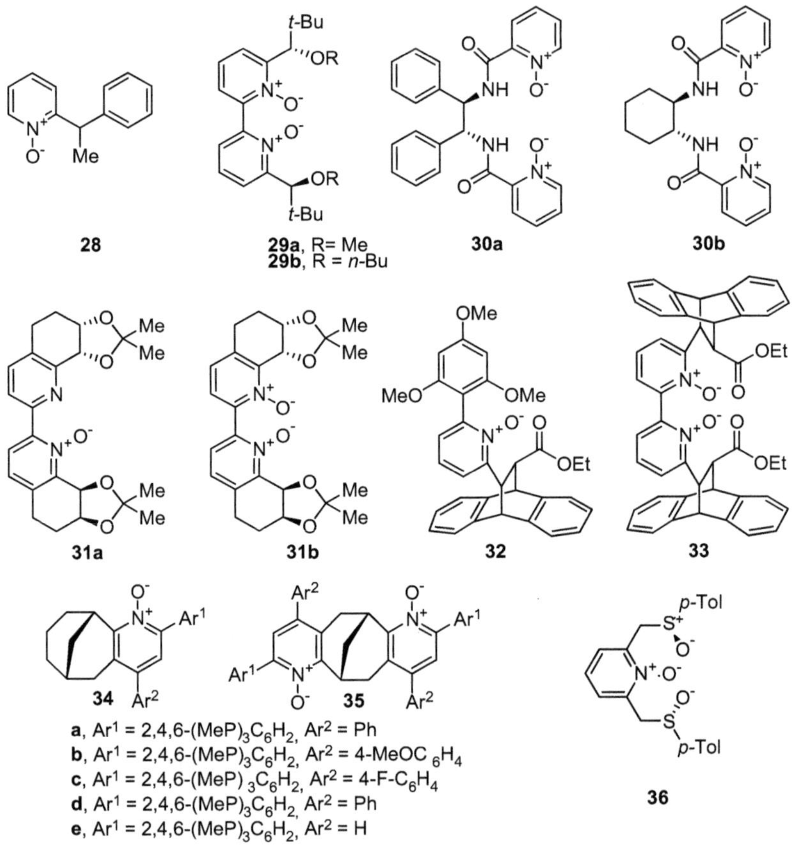

Fig. 6 Various pyridine *N*-oxides possessing elements of central chirality

were reported. Among these compounds axially chiral *N*-oxides **2** prepared by Kočovský et al. [45] and **3** prepared by Kotora et al. [46] could be also included.

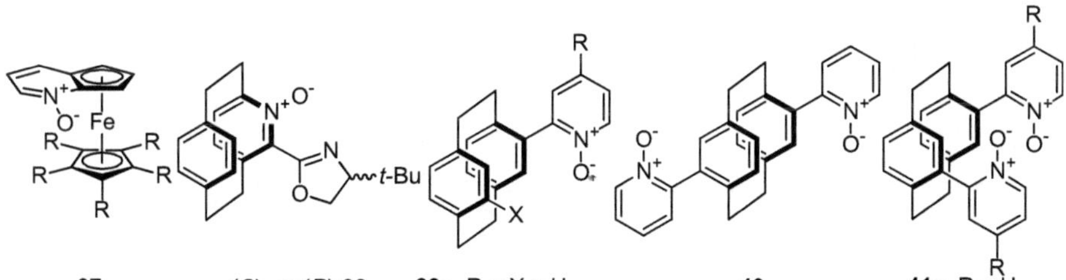

37 (*S*)- or (*R*)-**38** **39a**, R = X = H **39b**, R = OMe, X = H **39c**, R = NO_2, X = H **39d**, R = NO_2, X = Br **40** **41a**, R = H **41b**, R = OMe **41c**, R = NO_2

Fig. 7 Various pyridine *N*-oxides possessing elements of planar chirality

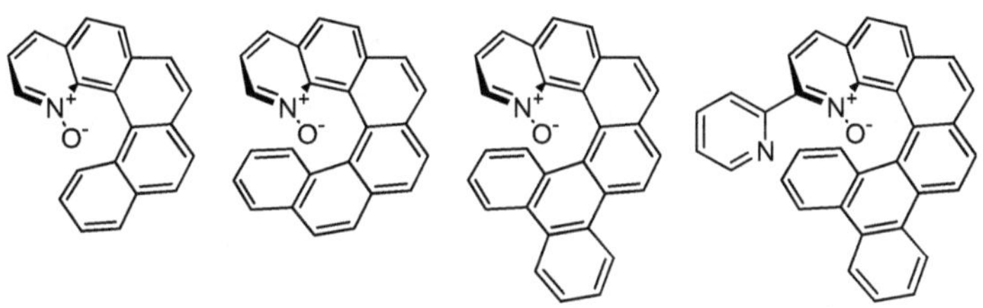

42 **43** **44** **45**

Fig. 8 Various pyridine *N*-oxides possessing elements of helical chirality

Further generations of unsymmetrically and symmetrically substituted chiral bipyridine *N,N′*-dioxides comprise **4a–4c**, and **5** possessing the tetrahydroisoquinoline framework, prepared by Kotora et al. [47–50], and, finally, *N,N′*-dioxides bearing the carbazole framework **6a** and **6b** prepared by Zhu et al. [51, 52]. Also a pyridine *N*-oxide with a chiral biphenyl moiety **7** prepared by Benaglia et al. [53] should be mentioned as well. A new type of catalyst **8** possessing the element of axial chirality was reported by Nakajima et al. [54] and a related compound also by Yu et al. [55].

Regarding the synthesis of bipyridine *N*-oxides and *N,N′*-dioxides possessing elements of central chirality, numerous compounds possessing different molecular scaffolds have been prepared (Figs. 4–6). Among them belong *N*-oxides derived from chiral natural terpenes (Fig. 4) such as bipyridine mono-*N*-oxides **9** [56], **10** (*iso*-PINDOX series) and **11** [57], structurally related pyridine *N*-oxides **12** [58] and **13** [59] prepared by Kočovský et al., **14** prepared by Benaglia et al. [60], **15** prepared by Denmark et al. [61], and, finally, **16** prepared by Marchetti et al. (absolute configuration was not given) [62]. Representatives of *N,N′*-dioxides are bipyridine *N,N′*-dioxide **17** and **18** prepared by Kočovský et al. [56, 57], bipyridine *N,N′*-dioxide **19** prepared by Denmark et al. [61], and bridged bipyridine *N,N′*-dioxides **20–22** prepared by Benaglia et al. [60]. Finally, *N,N′,N″*-trioxide **23** was prepared by Kwong et al. [63]. Among other *N*-oxides originating from natural source belongs a diastereomeric mixture of (1′*R*,2′*S*)- and (1′*S*, 2′*S*)-3-(1-methyl-2-pyrrolidinyl)pyridine *N,N′*-dioxides **24** synthesized by oxidation of (*S*)-(−)-3-(1-methyl-2-pyrrolidinyl)pyridine (nicotine) by Marchetti et al. [61].

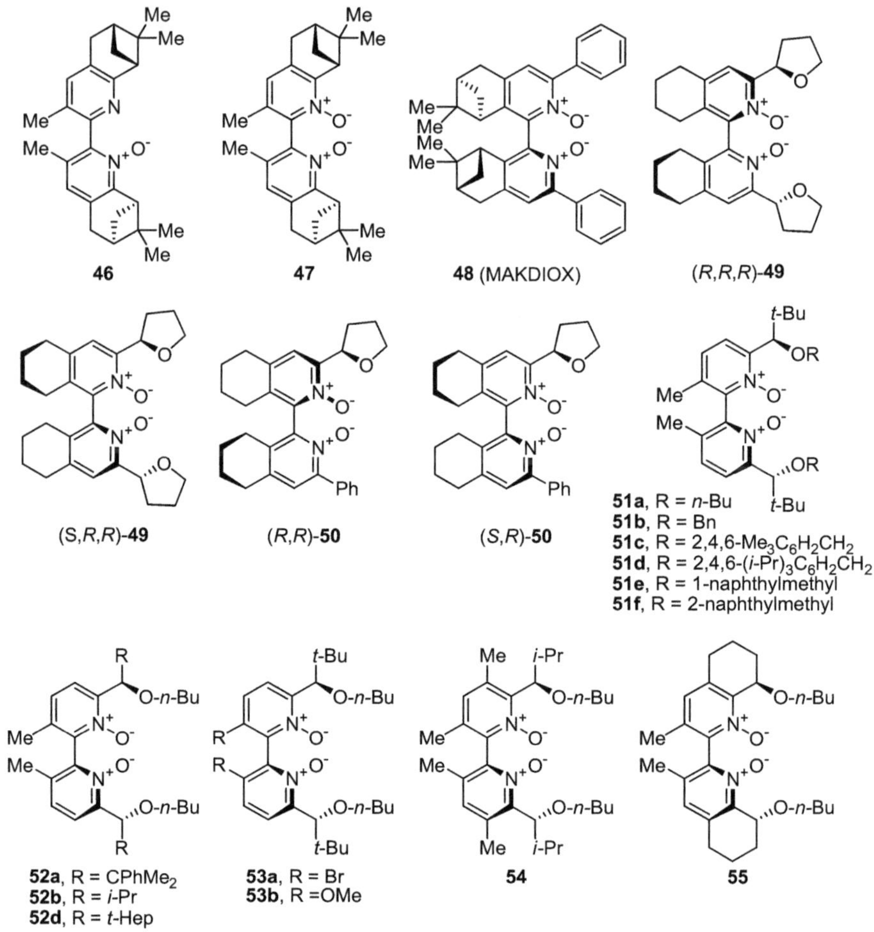

Fig. 9 Various pyridine *N*-oxides possessing elements of axial and central chirality

Another class of pyridine *N*-oxides possessing central chirality is derived from amino acids (Fig. 5). One example are pyridine *N*-oxides possessing amino acid as the part of the pendant chain **25** prepared by Benaglia et al. [64]. Other classes of pyridine *N*-oxides **26** and **27** with pendant amino acid chains were prepared by Laschat et al. [65].

The next group of *N*-oxides are those bearing the element of central chirality (Fig. 6). This group comprises various compounds such as pyridine *N*-oxide **28** prepared by Marchetti et al. (no absolute configuration was given) [60], Denmark et al. [66] reported a large series of chiral *N,N′*-dioxides **29** based on Bolm's catalysts scaffold [67, 68], *N,N′*-dioxides **30a** and **30b** prepared by Benaglia et al. [53], pyridine *N*-oxide **31a** and bipyridine *N,N′*-dioxide **31b** prepared by Boyd et al. [69], and finally *N*-oxides **32** and **33** were prepared by Ramanathan et al. [70, 71]. Stončius et al. [72] synthesized a series of *N*-oxides **34** and **35** in which the pyridine moieties were fused with the bicyclo[3.3.1]nonane framework. Finally, pyridine *N*-oxide **36** having additional bis-sulfoxide moieties was prepared by Juaristi et al. [73]. Worth mentioning are also amylose and cellulose derivatives bearing pyridine *N*-oxide groups [74].

Regarding *N*-oxides possessing an element of planar chirality, only a handful of compounds have been synthesized so far (Fig. 7). The pioneering work was done by Fu et al. [75] who prepared ferrocene containing pyridine *N*-oxide **37**. In addition, *N*-oxides possessing the [2.2]-paracyclophane scaffold **38** and **39–41** were prepared by Andrus et al. [76] and Rowlands et al. [77], respectively. A similar situation concerns *N*-oxides possessing elements of helical chirality (Fig. 8). A series of various helical *N*-oxides **42–44** [78, 79] and **45** [80] was prepared by Takenaka et al. (Fig. 8).

The last group (Fig. 9) comprise chiral bipyridine *N*-oxide **46** [56] and *N*,*N*′-dioxide **47** [57] having elements of both central and axial chirality that were prepared by Kočovský et al., *N*,*N*′-dioxide **48** was prepared by Malkov et al. [81], and finally *N*, *N*′-dioxides **49** and **50** bearing the chiral tetrahydrofuranyl moiety by Kotora et al. [49, 50]. A large series of variously substituted axially chiral bipyridine *N*-oxides **51–55** was prepared by Denmark et al. [66, 82]. This group also comprises other types of chiral *N*,*N*′-dioxides possessing similar structural features [83].

4.2 Allylations and Crotylations

N-Oxides of various types have been used as chiral Lewis bases able to activate Lewis acids (halosilanes), namely, in reactions of aldehydes with allyltrichlorosilane (Hosomi–Sakurai type allylations) providing chiral homoallylic alcohols. Their catalytic activity, i.e., increasing of the reaction rate and asymmetric induction, was usually tested in allylations of aromatic aldehydes. Allylation of benzaldehyde was usually chosen to assess the abovementioned properties. For typical examples, see the results summarized in Table 3. The outcome of the allylation usually depended on the amount of a catalyst and usually rather high loading such as 10 mol% with respect to the aldehyde and reaction times (24 or 48 h or more) were required to obtain a reasonable reaction rate and enantioselectivity. On the other hand, there exists a couple of *N*-oxides capable of catalyzing the reaction at a 1 mol% level providing the respective products in high yields and asymmetric induction within a reasonable period of time (1–6 h). Allylations of other benzaldehydes were studied as well and can be found elsewhere [84, 85].

A strong solvent effect on asymmetric induction as well as on the configuration of the homoallylic alcohol was observed during allylations catalyzed by **5**. Thus catalysis by (*R*)-**5** gave (*S*)-homoallylic alcohols in MeCN, CH_2Cl_2, $CHCl_3$, and other polar solvents; whereas its use in nonpolar solvents such as toluene, THF, chlorobenzene, etc. gave (*R*)-homoallylic alcohols [43]. Properties of (*S*)-**1d** were tested in allylation of 4-methoxybenzaldehyde only. It proceeded with a high enantioselectivity of 92% ee (66% yield) [43].

Another interesting and synthetically useful reaction is crotylation of aldehydes with (*E*)- and (*Z*)-crotyltrichlorosilanes. The former gives rise preferentially to *anti*

Table 3 Allylation of benzaldehyde catalyzed by various chiral *N*-oxides

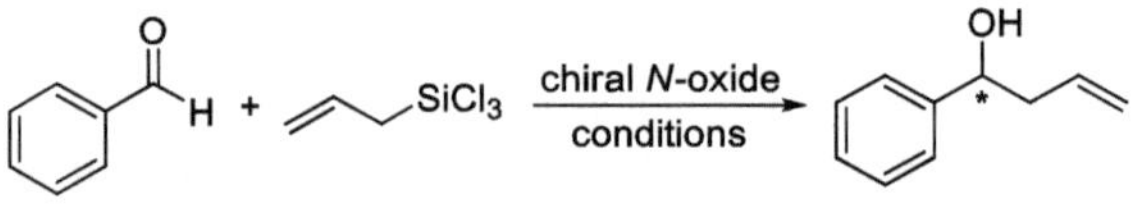

Catalyst	mol (%)	Solvent	Conditions	Yield (%)[a]	ee (%)[a]	Ref
(*S*)-**1a**	10	CH_2Cl_2	23°C, 2 h	82	52	[40]
(*S*)-**1b**	10	CH_2Cl_2	−78°C, 6 h	85	88 (*R*)	[40]
(*R*)-**1c**	0.1	CH_2Cl_2	−45°C, 2.5 h	95	84 (*S*)	[41]
(*R*)-**2**	5	CH_2Cl_2	−40°C, 2 h	60	87 (*R*)	[45]
(*R*)-**3**	5	CH_2Cl_2	20°C, 72 h	50	20	[46]
(*S*)-**4a**	5	CH_2Cl_2	−78°C, 6 h	87	74 (*R*)	[47]
(*S*)-**4b**	5	CH_2Cl_2	−78°C, 6 h	53	72 (*R*)	[47]
(*S*)-**4c**	1	CH_2Cl_2	−78°C, 1 h	48	80 (*R*)	[48]
(*R*)-**5**	1	Toluene[b]	−78°C, 1 h	45	83 (*S*)	[49]
(*R*)-**6a**	1	CH_2Cl_2	−80°C, 16 h	88	95 (*S*)	[51]
(*R*)-**6b**	1	CH_2Cl_2	−80°C, 20 h	–	87 (*R*)	[52]
(*R*)-**7**	10	MeCN	−90°C, 24 h	22	40 (*R*)	[53]
9	7	CH_2Cl_2	−90°C, 24 h	67	92 (*S*)	[56]
10a	20	CH_2Cl_2	−60°C, 18 h	72	46 (*S*)	[57]
10b	20	CH_2Cl_2	−60°C, 18 h	75	88 (*S*)	[57]
10c	20	CH_2Cl_2	−60°C, 18 h	72	84 (*S*)	[57]
10d	10	CH_2Cl_2	−60°C, 18 h	15	97 (*S*)	[57]
11	10	CH_2Cl_2	−60°C, 18 h	90	22 (*S*)	[57]
12a	7	MeCN	−60°C, 18 h	66	41 (*S*)	[58]
12b	7	MeCN	−60°C, 18 h	15	16 (*S*)	[58]
12c	7	MeCN	−60°C, 18 h	20	7 (*S*)	[58]
12d	7	MeCN	−60°C, 18 h	55	68 (*S*)	[58]
12e	7	MeCN	−60°C, 18 h	51	67 (*S*)	[58]
13	5	MeCN	−40°C, 18 h	95	96 (*S*)	[59]
14	10	MeCN	0°C, 48 h	17	6 (*S*)	[60]
17	7	CH_2Cl_2	−90°C, 48 h	18	41 (*R*)	[56]
20	10	MeCN	−40°C, 67 h	37	95 (*S*)	[60]
21a	10	MeCN	−40°C, 72 h	30	4 (*S*)	[60]
21b	10	MeCN	−40°C, 72 h	24	35 (*S*)	[60]
22	10	MeCN	−40°C, 72 h	22	30 (*S*)	[60]
23	10	CH_2Cl_2	0°C, 3 h	89	74 (*R*)	[63]
25a	10	MeCN	20°C, 48 h	85	0	[64]
25b	10	MeCN	20°C, 48 h	53	6	[64]
25c	10	MeCN	0°C, 48 h	45	68	[64]
25d	10	MeCN	−20°C, 48 h	40	67 (*S*)	[64]
25e	10	MeCN	0°C, 48 h	50	60 (*S*)	[64]
30a	10	MeCN	−45°C, 40 h	60	50 (*S*)	[53]
30b	10	MeCN	−45°C, 40 h	50	35 (*S*)	[53]
31a	10	CH_2Cl_2	0°C, 24 h	60	35 (*R*)	[69]

(continued)

Table 3 (continued)

Catalyst	mol (%)	Solvent	Conditions	Yield (%)[a]	ee (%)[a]	Ref
31b	10	CH_2Cl_2	−78°C, 12 h	64	26 (*R*)	[69]
32	20	$CHCl_3/(CH_2Cl)_2$	−78°C, 24 h	87	83 (*S*)	[70]
38	1.5	MeCN	−40°C, 6 h	95	93 (*S*)	[76]
39a	10	CH_2Cl_2	−78°C	65	38 (*R*)	[76]
39b	10	CH_2Cl_2	−78°C	52	36 (*S*)	[76]
41a	10	CH_2Cl_2	−78°C	72	28 (*S*)	[77]
41b	10	CH_2Cl_2	−78°C	58	30 (*S*)	[77]
46	10	CH_2Cl_2	−60°C, 12 h	72	98 (*S*)	[56]
47	10	CH_2Cl_2	−60°C, 12 h	52	14 (*S*)	[57]
(*R,R,R*)-**49**	1	MeCN	−40°C, 1 h	100	48 (*S*)	[49]
(*S,R,R*)-**49**	1	PhCl	−40°C, 1 h	100	62 (*S*)	[49]
(*R,R*)-**50**	1	THF	−78°C, 1 h	100	93 (*S*)	[50]
(*S,R*)-**50**	1	THF	−78°C, 1 h	98	96 (*S*)	[50]

[a]Configuration of the obtained alcohol is given in parentheses
[b]Allylations in other solvents were tested as well

whereas the latter affords *syn* products. Typical examples of benzaldehyde crotylations are displayed in Table 4. Crotylations of other substrates were reported as well [90]. A more detailed kinetic and computational investigation of the catalytic enantio- and diastereoselective allylation and crotylation is given elsewhere [91].

Allylation of α,β-unsaturated aldehydes has also been extensively studied. In the majority of the cases, only allylation of cinnamaldehyde and its substituted congeners was studied [40, 42, 45, 51, 56, 57, 60, 70, 81, 86, 88, 89, 92, 93]. More systematic studies were carried out in a handful of cases only [94–96]. Worth mentioning is also allylation (crotylation) of β-haloacrylaldehydes, products of which could be used as convenient chiral building blocks [97, 98]. Only a handful examples have been reported for allylations of α,β,γ,δ-unsaturated aldehydes and these mainly concerned reactions catalyzed by **12** [96], by (*R,R*) and (*S,R*)-**50** [97, 99, 100], or by (*R,R,R*)- and (*S,R,R*)-**49** [100]. Allylation of aliphatic aldehydes catalyzed by **5**, (*S,R,R*)-**49** [101] and (*S,R*)-**50** [102] was tested as well, but asymmetric induction was rather low and did not exceed 68% ee.

4.3 Propargylations and Allenylations

Selective propargylation and allenylation of aromatic aldehydes with propargyltrichlorosilane and allenyltrichlorosilane in the presence of 20 mol% of chiral *N,N′*-dioxide (S)-**1b** were reported by Denmark et al. [103]. Propargyltrichlorosilane and allenyltrichlorosilane were prepared in situ by a reaction of trichlorosilane with propargyl chloride either under Cu or Ni catalysis, respectively. Although it was the first example,

Table 4 Crotylations of benzaldehyde catalyzed by various chiral *N*-oxides

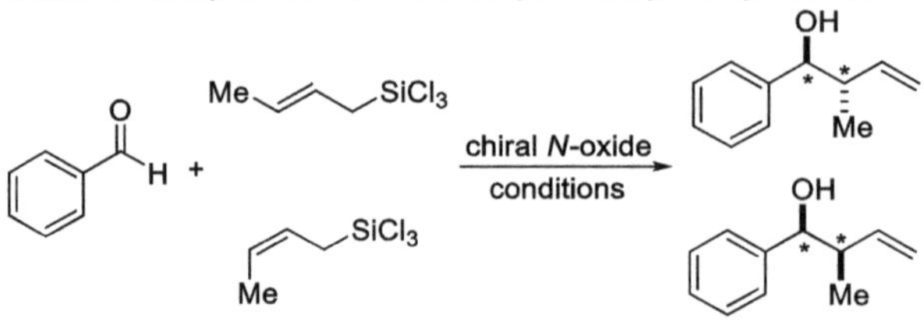

Catalyst	mol (%)	*trans* or *cis*	Solvent	Conditions	Yield (%)[a]	*anti*/*syn*	ee (%)[a] *anti*, *syn*	Ref
(*S*)-**1b**	10	trans	CH_2Cl_2	−78°C, 6 h	68	97:3	86 (1*R*,2*R*)	[40]
(*S*)-**1b**	10	cis	CH_2Cl_2	−78°C, 6 h	64	1:99	84 (1*R*,2*S*)	[40]
(*S*)-**1c**	10	trans	MeCN	−45°C, 2.5 h	84	96:4	73 (1*S*,2*S*)	[86]
(*S*)-**1c**	10	cis	MeCN	−45°C, 2.5 h	82	1:99	77 (1*S*,2*R*)	[86]
(*R*)-**2**	5	trans	CH_2Cl_2	−40°C	54	93:7	87	[45, 87]
(*S*)-**2**	5	trans	CH_2Cl_2	−40°C, 24 h	65	95:5	66 (1*S*,2*S*), 77 (1*S*,2*R*)	[88]
(*S*)-**2**	5	cis	CH_2Cl_2	−40°C, 24 h	78	1:99	69 (1*S*,2*S*), 79 (1*S*,2*R*)	[88]
(*R*)-**6a**	1	trans	CH_2Cl_2	−80°C, 16 h	64	86:14	94 (1*S*,2*S*), 95	[89]
9	7	trans	CH_2Cl_2	−40°C	70	68:32	65/78	[45]
(−)-**10d**	10	trans	MeCN	−40°C, 18 h	88	98:2	98	[57]
(−)-**10d**	10	cis	MeCN	−40°C, 18 h	37	10:90	87	[57]
25c	1	trans/cis, 8/2	MeCN	0°C, 48 h	40	80:20	69	[64]
(*R*,*R*)-**50**	1	trans	PhCl	−40°C, 24 h	82	71:29	91/87	[50]

[a]Configuration of the obtained alcohol is given in parentheses

a high catalyst loading, moderate yields, and enantioselectivity were short of expectations (Table 5).

Later, Takenaka et al. developed a highly enantioselective synthesis of homopropargylic alcohols by using a new helical chiral *N*-oxide catalyst **45** and allenyltrichlorosilane (Table 6) [80, 104]. Reactions with various aldehydes provided the corresponding homopropargylic alcohols in high yields and with high enantioselectivity, when 10 mol% catalyst loading was used. The reaction tolerates functionalities in the arene, such as halogen, nitro, trifluoromethyl, and methoxy groups. However, the reaction of cyclohexylcarbaldehyde afforded the product in a lower yield and with a moderate enantioselectivity. In addition, this method was used for enantio- and regioselective propargylation of *N*-acylhydrazones (Scheme 4).

Table 5 Selective synthesis of optically active allenyl and homopropargyl alcohols

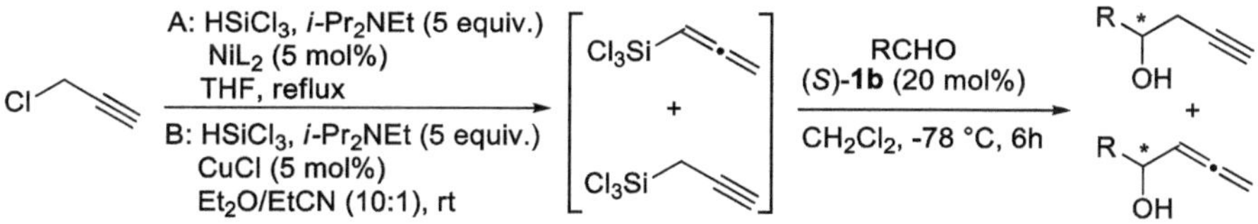

Entry	Method	Aldehyde	Yield (%)[a]	Propargyl:allenyl[b]	ee (%)[c, d]	Configuration[e]
1	A	Ph	65	>30:1	52	*R*
2	A	4-$MeOC_6H_4$	62	>30:1	40	*R*
3	A	4-ClC_6H_4	49	>30:1	46	*R*
4	A	$Ph(CH_2)_2$	35	>30:1	23	*R*[f]
5	B	Ph	72	1:15	54	*R*
6	B	4-$MeOC_6H_4$	76	1:9	62	*R*[f]
7	B	4-ClC_6H_4	48	1:9	49	*R*[f]
8	B	$Ph(CH_2)_2$	44	1:10	22	*R*[f]

[a]Combined isolated yields
[b]Determined by ^{1}H NMR
[c]Enantiomeric excess of the major isomer
[d]Determined by HPLC
[e]Assigned by comparison with optical rotation and/or retention time on chiral HPLC
[f]Assigned by analogy

The initial idea on the supposed origin of high enantioselectivity based on π–π-stacking between the phenyl ring of the aromatic aldehyde and the helical core of the catalyst was corrected later by the theoretical study of Wheeler and coworkers. They reported a computational study showing that simple electrostatic interactions stabilize the preferred *Si* face addition [105].

4.4 *Aldol Reactions*

4.4.1 Aldol Reaction with Ketones

Compared to the multitude of Lewis acid-catalyzed enantioselective aldol additions, the Lewis base catalysis is considerably scarce. Concerning the Lewis base catalysts, chiral phosphoramides introduced by Denmark et al. have been studied most extensively. In 2002, he reported the application of chiral *N*-oxide catalysis for aldol addition to unactivated ketones [61, 66], which, unlike an addition to aldehydes, is a considerably more challenging reaction, and therefore a general solution was lacking. Using highly reactive trichlorosilyl ketene acetals[1] enabled to overcome the diminished electrophilicity and increased steric hindrance of ketones compared to aldehydes. It is

[1]Trichlorosilyl enolates of aldehydes and ketones as the reagents of the first choice were found not reactive enough for addition to ketones.

Table 6 Asymmetric propargylation of aldehydes catalyzed by helically chiral *N*-oxide **45**

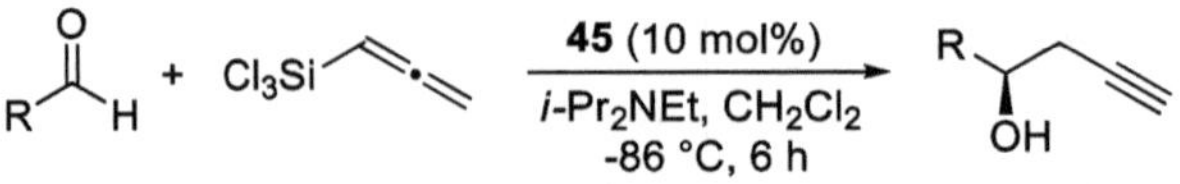

Entry	R	Yield (%)[a]	ee (%)[b]
1[c]	Ph	87	86
2[c]	2-Naphthyl	86	84
3[c]	4-Br-C_6H_4	95	92
4	4-Cl-C_6H_4	90	92
5	4-F-C_6H_4	93	88
6[d]	4-NO_2-C_6H_4	55[e]	92
7	4-CF_3-C_6H_4	80	90
8	4-MeO-C_6H_4	80	74
9	4-Me-C_6H_4	85	82
10[c]	2-Br-C_6H_4	93	96
11	2-Cl-C_6H_4	97	96
12	2-F-C_6H_4	98	92
13	2-NO_2-C_6H_4	87	96
14[d]	2-CF_3-C_6H_4	95	94
15	2-MeO-C_6H_4	78	94
16	2-Me-C_6H_4	90	86
17[d]	2-Br-4-Me-C_6H_4	92	96
18[f]	Cy	61[g](80)[h]	59

[a]Isolated yield
[b]Determined by HPLC
[c]Absolute configurations were determined
[d](*M*)-**43** catalyst was used
[e]Low yield is presumably due to a poor solubility of the aldehyde
[f](*R*)-isomer is major
[g]12 h
[h]36 h

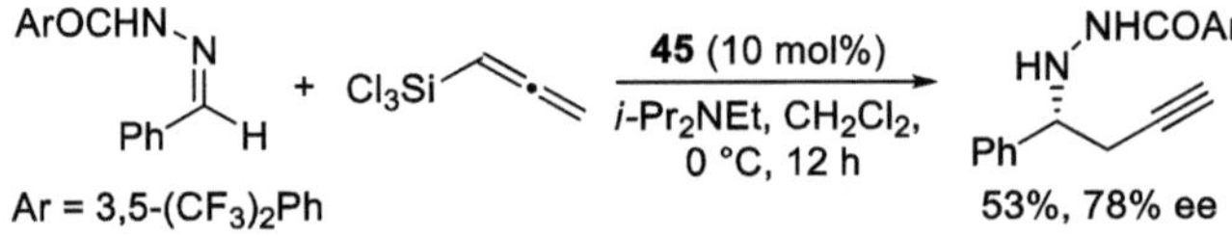

Scheme 4 Propargylation of an acylhydrazone catalyzed by **45**

worth mentioning that enantioselectivities reached with chiral pyridine *N*-oxides were superior to those obtained with the aforementioned phosphoramides. Vast effort was undertaken to optimize properties of the *N*-oxide catalysts (Table 7).

It was concluded that the asymmetric inductions and yields of the aldol product obtained with at-that-time-known monomeric **15** and dimeric axially chiral **19** or centrally chiral catalysts (*R*)-**1a** and (*S*)-**1g** are overcome by pyridine *N*-oxides **29a** and

Table 7 Enantioselective aldol addition to ketone – *N*-oxide catalysts optimization

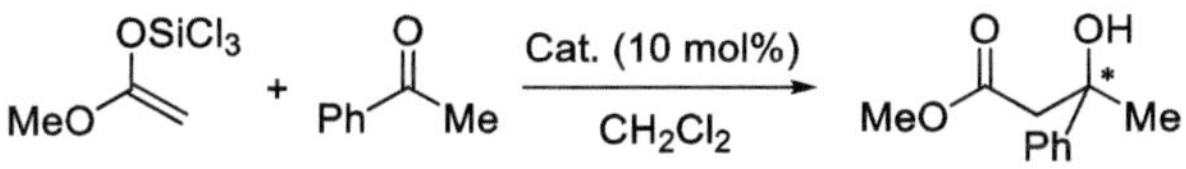

Entry	Catalyst	Temp. (°C)	Yield (%)	ee (%)	References
1	**15**	0	55	3	[60]
2	**19**	0	15	<1	[60]
3	(*R*)-**1a**	0	45	45	[65]
4	(*S*)-**1g**	0	39	26	[65]
5	**29a**	0	92	55	[65]
6	**29b**	0	94	64	[65]
7	(*P*)-**51a**	0	90	74 (*S*)	[65]
8	(*P*)-**51a**	−20	94	82 (*S*)	[65]
9	(*M*)-**51a**	−20	89	42 (*R*)	[65]
10	**51b**	−20	89	82 (*S*)	[60]
11	**51c**	−20	91	87 (*S*)	[60]
12	**51d**	−20	64	60 (*S*)	[60]
13	**51e**	−20	88	74 (*S*)	[60]
14	**51f**	−20	91	1 (*R*)	[60]
15	**52b**	−20	93	64 (*S*)	[60]
16	**52a**	−20	90	80 (*S*)	[60]
17	**52c**	−20	87	47 (*S*)	[60]
18	**53a**	−20	32	61 (*S*)	[60]
19	**53b**	−20	95	72 (*S*)	[60]
20	**54**	−20	93	72 (*S*)	[60]
21	**55**	−20	92	24 (*S*)	[60]
22	**55**	−20	94	5 (*R*)	[60]

29b derived from the Bolm's ligand. A significant improvement of enantioselectivity was reached by blocking the rotation along the C2–C2′ bond by installing methyl groups in positions 3 and 3′. Thus the presence of the enantioselectivity-attenuating axial conformer of the catalyst was avoided, and catalysts **51** and **52** demonstrated the synergy of central and axial chirality (compare Entries 8 and 9 with Entry 7). Regarding the steric influences, the best selectivity was obtained with catalyst **51c** possessing *tert*-butyl and 2,4,6-trimethylbenzyl groups. With respect to the electronic influences, the decreased yield was obtained with catalyst **53a** clearly indicating that electron deficient pyridine *N*-oxides (weaker Lewis bases) are less capable of catalyzing the reaction.

The study aiming to determine the scope of the reaction revealed that the greater the difference in R^1 and R^2 bulkiness is, the higher enantioselectivities are obtained (Table 8). Furthermore, aldol condensation with aromatic ketones (Entries 1–9) resulted in higher enantioselectivities than with the aliphatic ones, whether linear, cyclic, branched, or conjugated (Entries 10–16).

Table 8 The scope of the pyridine *N*-oxide catalyzed aldol addition with respect to various ketones

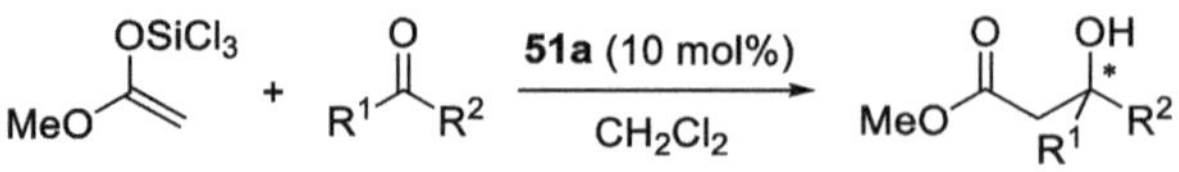

Entry	Ketone	Yield (%)	ee (%)
1	PhCOMe	96	83
2	PhCOEt	90	81
3	PhCOC≡CH	89	86
4	1-Tetralone	90	80
5	1-NaphtylCOMe	89	56
6	2-FurylCOMe	87	49
7	PhC≡CCOMe	94	35
8	*p*-$CF_3C_6H_4COMe$	91	76
9	*p*-$MeOC_6H_4COMe$	94	68
10	(*E*)-PhCH=CHCOMe	87	11
11	2-Cyclohexenone	86	8
12	EtCOMe	84	32
13	$PhCH_2CH_2COMe$	97	35
14	cyclopropylCOMe	84	20
15	cyclohexylCOMe	91	32
16	*t*-BuCOMe	87	43

4.4.2 Aldol Reaction with Aldehydes

Nakajima et al. noticed that addition of trichlorosilyl enolates is analogous to the previously studied additions of allyltrichlorosilane to aldehydes and therefore a similar mechanism and enantioselectivities to those of the allylation reactions could be expected in the presence of *N*-oxide catalysts (Table 9) [54]. It was discovered that: (a) the aldolizations are high-yielding except for aliphatic aldehyde (Entry 11); (b) unlike Lewis acid-catalyzed reactions it is stereospecific, turning *E* enolates into *syn* aldol products and *Z* enolates into *anti* ones (Entries 3–10), and (c) the asymmetric induction in the presence of catalysts (*R*)-**1a** or (*R*)-**1g** strongly depended on the substrate and catalyst structure.

In the case of trichlorosilyl enolates of cyclic ketones, modest enantioselectivities were observed, when bidentate catalysts (*R*)-**1a** and (*R*)-**1g** were used (Table 10); nevertheless, the diastereoselectivity followed the previous pattern yielding predominantly the *anti*-products [54]. However, employing monodentate catalysts (*R*)-**2** and (*R*)-**8** led to isolation of *syn* products with somewhat higher enantioselectivity.

Table 9 Aldol addition of trichlorosilyl enolates of acyclic ketones to aldehydes catalyzed by **1**

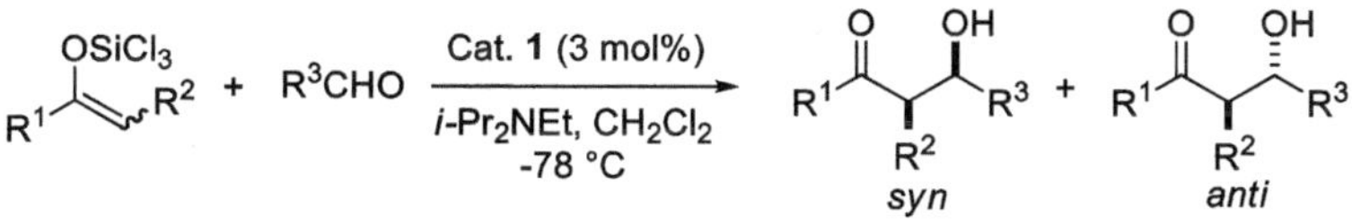

Entry	R^1	R^2	*E/Z*	R^3	Catalyst	Yield (%)	*syn/anti*	ee (%) *syn, anti*
1	Ph	H	–	Ph	(*R*)-**1g**	85	–	<5
2	Ph	H	–	Ph	(*R*)-**1a**	87	–	<5
3	H	C_5H_{11}	12/1	Ph	(*R*)-**1g**	96	1/11	6, 7
4	H	C_5H_{11}	12/1	Ph	(*R*)-**1a**	86	1/12	81, 23
5	H	C_5H_{11}	1/4	Ph	(*R*)-**1g**	88	3/1	9, 12
6	H	C_5H_{11}	1/4	Ph	(*R*)-**1a**	90	4/1	79, 23
7	Ph	Me	1/10	Ph	(*R*)-**1g**	82	7/1	82, 33
8	Ph	Me	1/10	Ph	(*R*)-**1a**	88	9/1	6, <5
9	Ph	Me	1/10	4-MeOC_6H_4	(*R*)-**1g**	86	15/1	67, 13
10	Ph	Me	1/10	PhCH=CH	(*R*)-**1g**	59	15/1	63, 43
11	Ph	Me	1/10	$PhCH_2CH_2$	(*R*)-**1g**	Trace	–	–

Table 10 Aldol addition of trichlorosilyl enol ethers of cyclic ketones to aldehydes

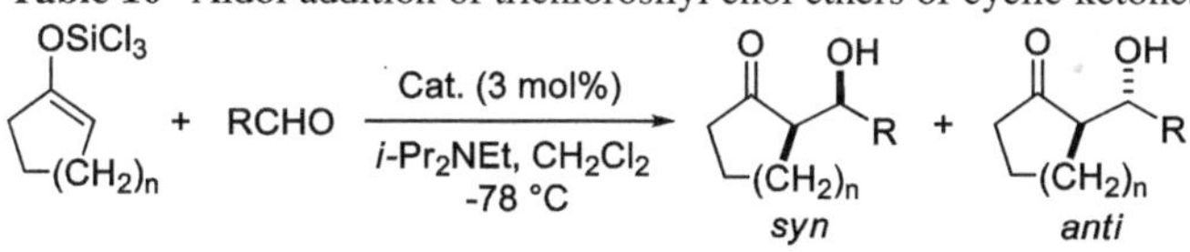

Entry	*n*	R^3	Catalyst	Yield (%)	*syn/anti*	ee (%) *syn, anti*
1	2	Ph	(*R*)-**1g**	80	1/10	39, 30
2	2	Ph	(*R*)-**1a**	94	1/3	21, 30
3	2	Ph	(*R*)-**2**	92	8/1	<5, 30
4	2	Ph	(*R*)-**8**	92	25/1	47, 60
5	3	Ph	(*R*)-**8**	93	30/1	50, 11
6	1	Ph	(*R*)-**8**	94	13/1	62, 66
7	1	4-MeOC_6H_4	(*R*)-**8**	90	14/1	63, 55
8	1	$4\text{-CF}_3C_6H_4$	(*R*)-**8**	98	14/1	72, 69
9	1	PhCH=CH	(*R*)-**8**	93	4/1	28, 42
10	1	$PhCH_2CH_2$	(*R*)-**8**	22	1/1	50, 40

4.4.3 Reductive Aldol Addition

Nakajima et al. also reported a highly *syn*-selective reductive aldol reaction of an α,β-unsaturated ketone – chalcone – with benzaldehyde [106]. 1,4-Reduction of the ketone by trichlorosilane in the presence of a pyridine *N*-oxide catalyst gave rise selectively to a (*Z*)-trichlorosilyl enolate. Once the highly pure (*Z*)-enolate was formed (due to the cyclic transition state), the *N*-oxide catalyst facilitated the enantioselective addition to a carbonyl acceptor. The use of (*R*)-**1a** and (*R*)-**1g** catalysts gave rise to the aldol products with comparably high diastereoselectivity (~19/1), while a satisfactory asymmetric induction of 80% ee was obtained only in the case of the former catalyst (Scheme 5). An intramolecular reaction was reported as well, the yield and selectivity were moderate (Scheme 6).

4.5 Alkylations

There is only one example of *N*-oxide catalyzed alkylation of aldehydes and it was reported by Laschat et al. [65]. They used pyridine *N*-oxides **26** and **27** possessing amino acid moieties as catalysts for alkylations of benzaldehyde with diethylzinc (Table 11). The reaction proceeded with good yields of the homoallyl alcohols in the range of 37–92%, but with a rather poor enantioselectivity in the range of 2–29% ee.

4.6 Epoxide Cleavage

Epoxide cleavage with various electrophiles leads to the formation of chiral alcohols. The selected examples are shown in Table 12. The first chiral Lewis base catalyzed epoxide ring cleavage was reported by Fu et al. in 2001. He studied cleavage of symmetrically substituted epoxides with tetrachlorosilane catalyzed by a ferrocene containing pyridine *N*-oxide **37**. The reaction proceeded usually with a high enantioselectivity (91–98% ee) for aryl substituted epoxides [75]. Takenaka et al. studied catalytic activity of several helically chiral *N*-oxides **42–44** [78]. Out of these, the use of **44** gave products

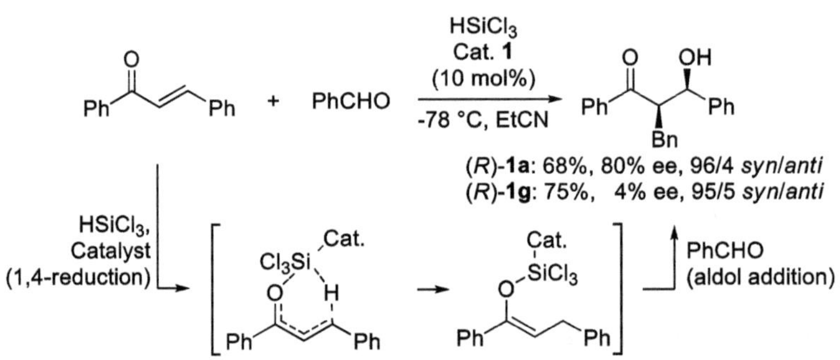

Scheme 5 Intermolecular enantioselective tandem 1,4-reduction-aldol additions catalyzed by (*R*)-**1**

Scheme 6 An intramolecular enantioselective tandem 1,4-reduction-aldol addition catalyzed by (*R*)-**1g**

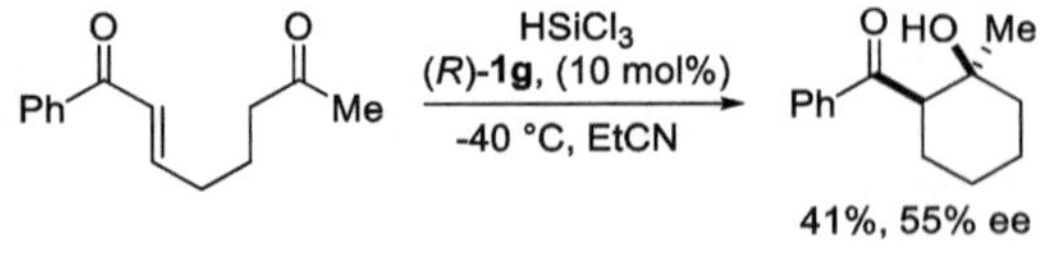

Table 11 Alkylation of benzaldehyde catalyzed by **26** and **27**

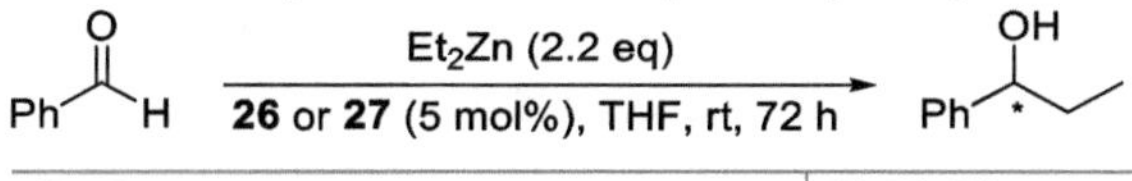

Catalyst	Yield (%)	ee (%)[a]
26a	52	16 (*S*)
26b	37	12 (*S*)
26c	37	2 (*S*)
26d	71	9 (*S*)
27a	71	11 (*S*)
27b	88	7 (*S*)
27c	92	29 (*S*)
27d	88	8 (*R*)

[a]Determined by GC

Table 12 Ring opening of symmetrically substituted epoxides by various *N*-oxide catalysts

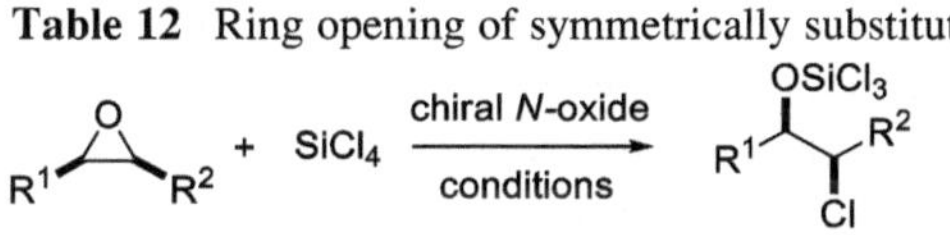

Catalyst	mol (%)	R^1, R^2	Solvent	Conditions	Yield (%)	ee (%)[a]
37	5	Ph	CH_2Cl_2	−85°C	88	94
37	5	4-FC_6H_4	CH_2Cl_2	−85°C	97	91
37	5	4-MeC_6H_4	CH_2Cl_2	−85°C	94	93
37	5	4-$CF_3C_6H_4$	CH_2Cl_2	−85°C	93	98
37	5	2-naphtyl	CH_2Cl_2	−85°C	84	94
37	5	CH_2OBn	CH_2Cl_2	−85°C	91	50
44	10	Ph	CH_2Cl_2	−78°C, 6 h	77	94
44	10	2-naphtyl	CH_2Cl_2	−78°C, 6 h	84	92
44	10	4-ClC_6H_4	CH_2Cl_2	−78°C, 6 h	83	94
44	10	4-MeC_6H_4	CH_2Cl_2	−78°C, 6 h	83	92
44	10	4-$CF_3C_6H_4$	CH_2Cl_2	−78°C, 6 h	63	87
44	10	$CH_2O(CH_2)_3Ph$	CH_2Cl_2	−78°C, 6 h	64	72
33	0.5	Ph	$CHCl_3$	−30°C, 2 h	94	93 (1*R*,2*R*)
33	0.5	4-FC_6H_4	$CHCl_3$	−30°C, 1.7 h	92	89 (1*S*,2*S*)
33	0.5	4-MeC_6H_4	$CHCl_3$	−30°C, 3 h	97	78 (1*S*,2*S*)
33	0.5	3-$MeOC_6H_4$	$CHCl_3$	−30°C, 1.5 h	96	89(1*R*,2*R*)
(*R*,*R*,*R*)-**49**	5	Ph	THF	−78°C, 24 h	76	65
(*S*,*R*,*R*)-**49**	5	Ph	THF	−78°C, 24 h	74	69

[a]Determined by GC

with the highest asymmetric induction. Another contribution to this area comes from Ramanathan et al. [71], who used conformationally rigid bipyridine *N*,*N*′-dioxide **33** as the catalyst. Comparison of catalytic activity of **49** and **50** revealed that the former gave better results (higher enantioselectivity) than the latter one, albeit its activity could not match the previously published results [107]. In a similar manner, *N*-oxides with the bicyclo[3.3.1]nonane framework gave in general products with low enantiopurity [72].

4.7 α-Addition of Isocyanides to Aldehydes (Passerini-Type Reaction)

In principle, the Passerini reaction could be also carried out under Lewis acid catalysis, but several problems were recognized in individual steps. However, these problems could be circumvented by the use of the concept of the Lewis base activation of Lewis acids. The main feature of this reaction is the activation of a weak Lewis acid, $SiCl_4$, by a weak chiral Lewis base generating a highly reactive and selective siliconium ion that reacts with an aldehyde. The ensuing reaction steps then give rise to an imidoyl chloride that after hydrolysis forms the expected product – a hydroxy amide. It has been shown that this reaction could be catalyzed by chiral bisphosphoramides to give highly enantioenriched products. Also *N,N′*-dioxide (*R,R,R*)-**51a** possessing elements of central and axial chirality was tested, but, disappointingly, its use did not show any sensible enantioselectivity. Thus the reaction of *tert*-butyl isocyanide **2**, benzaldehyde **3**, and tetrachlorosilane catalyzed by **51a** gave the corresponding α-hydroxy amide in a very good yield of 87% but with the poor enantioselectivity of 3% ee (Scheme 7) [29].

4.8 Addition of Me_3Si-CN to Ketones and Related Compounds (Strecker Reaction)

Addition of trimethylsilyl cyanide to acetophenone catalyzed by dual Lewis acid/Lewis base system composed of (*R*)-BINOL-Ti(O*i*-Pr)$_4$/(*R*)-**1b** was studied as a part of mechanistic studies by Feng et al. [108]. Comparison of the reaction carried out in the presence of racemic and chiral Lewis bases showed differences in enantioselectivity (43 and 51% ee) indicating a positive effect of the chiral Lewis base on asymmetric induction (Scheme 8). Addition of trimethylsilyl cyanide to aldimines promoted by various chiral *N*-oxides was studied as well [109] (for theoretical studies, see: Su et al.

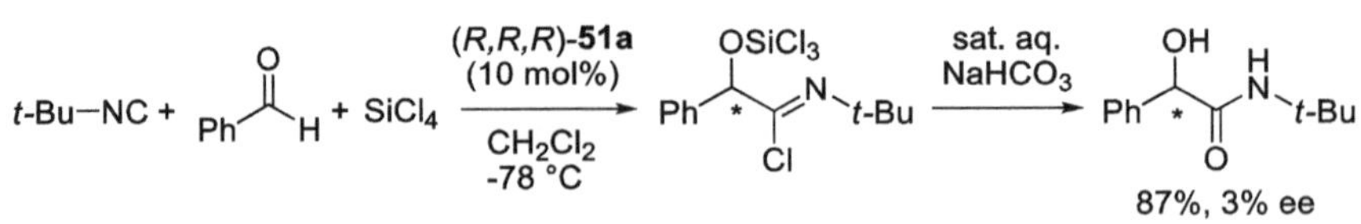

Scheme 7 Passerini-type reaction catalyzed by **51a**

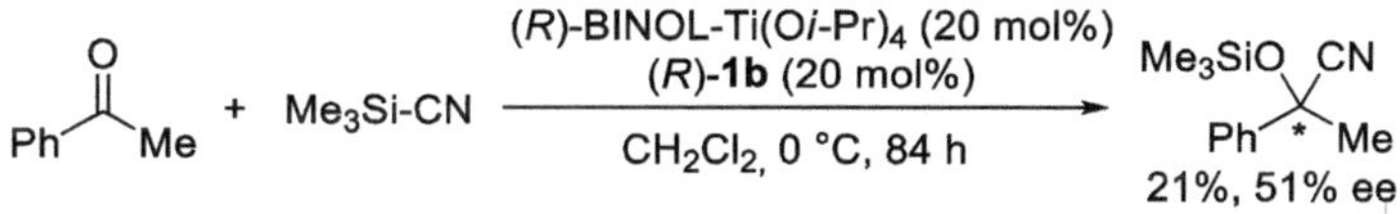

Scheme 8 Trimethylsilyl cyanide addition to acetophenone catalyzed by (*R*)-**1b**

[110]). However, in order to achieve a high asymmetric induction (up to 95% ee), the presence of stoichiometric amount of the chiral *N*-oxide was required.

4.9 Reductions

Reductions catalyzed by chiral pyridine *N*-oxide are rather rare. There is one example reported by Nakajima et al., who studied reduction of *N*-acylated β-amino enones (Scheme 9) [111]. Reduction of *N*-benzoyl enone by $HSiCl_3$ catalyzed by (*R*)-**1b** gave a mixture of 4*H*-1,3-oxazine (20%) and keto amide (18%) but with a rather low enantioselectivity of 37 and 53% ee, respectively.

The second example of reduction catalyzed by chiral pyridine *N*-oxide was reported by Laschat et al., who reported the reduction of ketones by $BH_3{\cdot}SMe_2$ in the presence of pyridine *N*-oxides **26** or **27** [65]. The reductions were carried out with just three ketones. The corresponding alcohols were obtained in very good yields, but with a low or moderate enantioselectivity, in the range of 7–64% (Table 13). The highest enantioselectivity (64% ee) was observed in the reduction of α-chloroacetophenone catalyzed by the monosubstituted catalyst **26d** (Entry 4).

4.10 Rearrangements

There is only one report on a rearrangement catalyzed by a chiral pyridine *N*-oxide. It concerns the synthesis of thiols via a rearrangement of carbonodithioic *O*,*S*-esters to carbonodithioic *S*,*S*-esters followed by transformation into thiols [112]. The rearrangement was studied in the presence of a terpene based pyridine *N*-oxide **16** and nicotine derived *N*,*N'*-dioxide **24** (Scheme 10). The use of the former did not result in any measurable asymmetric induction and the latter provides the corresponding thiols with a rather low enantioselectivity of 37.7% ee for the butane-2-thiol and 11.4% ee for the 1-phenylethylthiol.

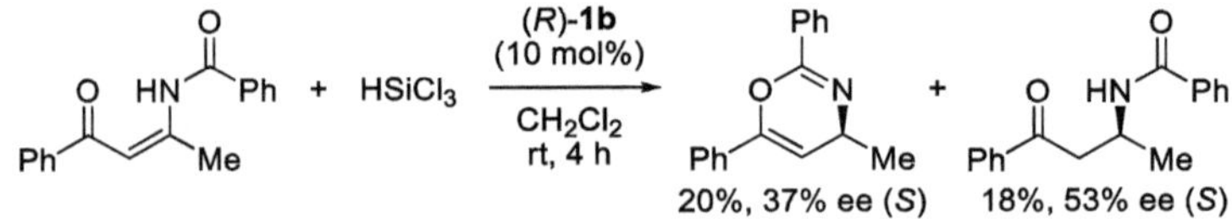

Scheme 9 Reduction of *N*-benzoyl enone by (*R*)-**1b**

Table 13 Reductions of ketones catalyzed by **26** and **27**

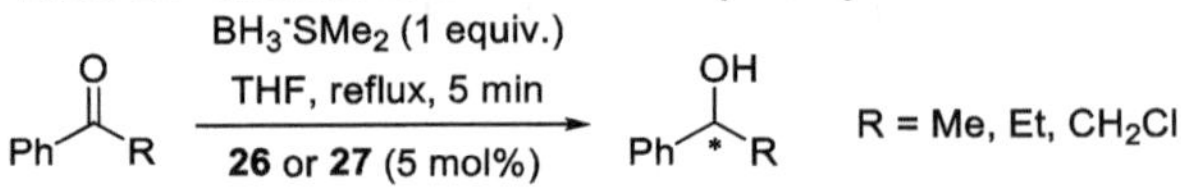

		R = Me		R = Et		R = CH_2Cl	
Entry	Catalyst	Yield (%)[a]	ee (%)[b]	Yield (%)[a]	ee (%)[b]	Yield (%)[a]	ee (%)[b]
1	**26a**	99	7 (*R*)	98	8 (*R*)	99	20 (*S*)
2	**26b**	99	11 (*R*)	99	17 (*R*)	97	21 (*S*)
3	**26c**	99	7 (*R*)	99	12 (*R*)	99	16 (*S*)
4	**26d**	99	20 (*R*)	98	21 (*R*)	99	64 (*S*)
5	**27a**	97	9 (*R*)	98	9 (*R*)	99	21 (*S*)
6	**27b**	99	10 (*R*)	99	16 (*R*)	99	33 (*S*)
7	**27c**	99	32 (*R*)	99	38 (*R*)	99	37 (*S*)
8	**27d**	99	31 (*R*)	99	33 (*R*)	98	51 (*S*)

[a]Isolated yields
[b]Determined by capillary GC

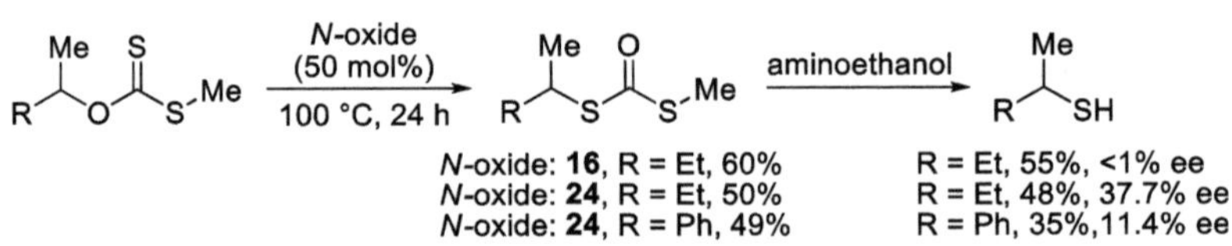

Scheme 10 Rearrangement of carbonodithioic *O*,*S*-esters to carbonodithioic *S*,*S*-esters by **16** and **24** (ee's were determined by optical rotation)

5 *N*-Oxide Catalyzed Reactions in Syntheses of Natural Compounds

Out of a huge number of chiral *N*-oxides, only a handful of them was used as catalysts for preparation of chiral building blocks that served as advanced intermediates for syntheses of natural or biologically active substances. Nonetheless several published examples nicely demonstrate their synthetic potential.

5.1 Naturally Occurring Compounds

Several chiral *N,N′*-dioxides were used to catalyze processes that led to the formation of chiral intermediates applied in syntheses of natural products. Interestingly, most of them were applied in enantioselective allylation of various unsaturated aldehydes. Thus, allylation of cinnamaldehyde with allyltrichlorosilane catalyzed by (*S,R*)-**50** was used as a crucial step in the synthesis of goniothalamin [94] (Scheme 11). In a similar manner, **48** was used to catalyze a highly enantioselective *syn* crotylation of substituted cinnamaldehydes with (*Z*)-crotyltrichlorosilane. The crotylation products were used to synthesize (−)-elisabethadione [81] and (−)-erogorgiaene [93] (Schemes 12 and 13). Allylation of an α,β,γ,δ-dienal with allyltrichlorosilane catalyzed by (*S,R*)-**50** was used to prepare the known left-hand fragment of papulacandin D (Scheme 14) [100]. The same approach was used for the total synthesis of (+)-pteroenone that was based on *anti*-crotylation of 2,4-dimethylhexa-2,4-dienal with (*E*)-crotyltrichlorosilane (Scheme 15) [99]. Alternatively, *anti*-crotylation of (*E*)-3-iodomethacryldehyde was

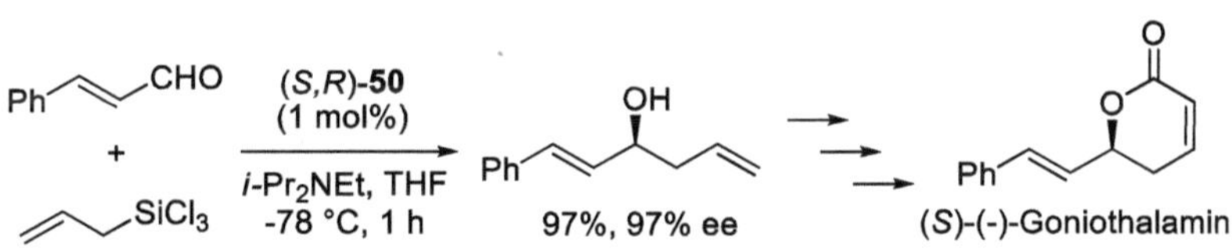

Scheme 11 Synthesis of (*S*)-(−)-goniothalamin

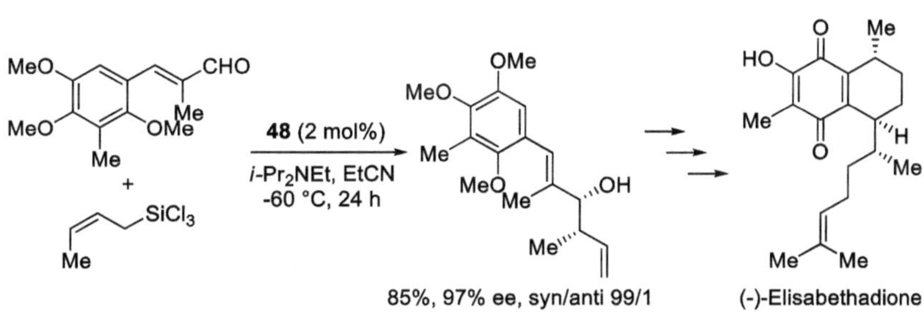

Scheme 12 Synthesis of (−)-elisabethadione

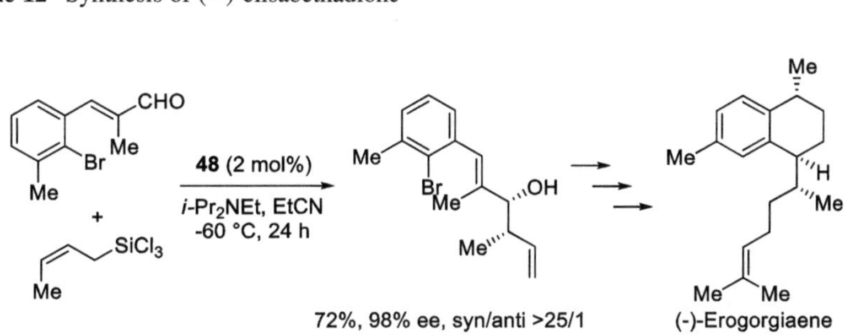

Scheme 13 Synthesis of (−)-erogorgiaene

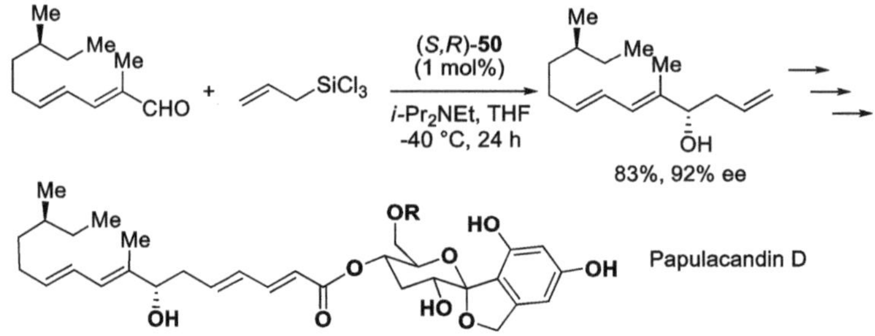

Scheme 14 Synthesis of the left-hand fragment of papulacandin D

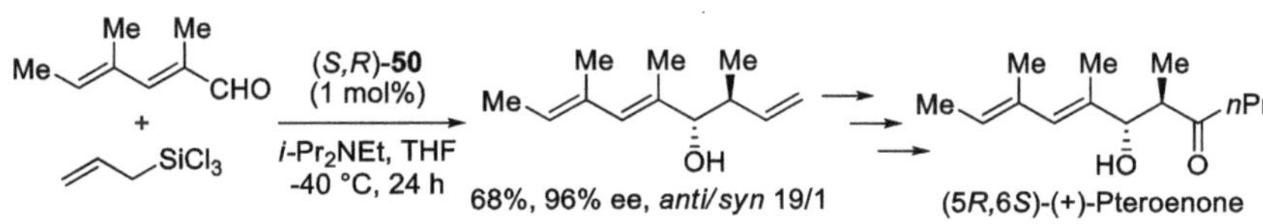

Scheme 15 Synthesis of pteroenone

used to provide the advanced intermediate for the syntheses of pteroenone and antillatoxin (Scheme 16) [97]. Enantioselective allylation was also exploited in the synthesis of one of the chiral building blocks utilized in the synthesis of the callyspongiolide fragment [113]. Allylation of 6-heptenal catalyzed by (*S*,*R*)-**50** was tested as a route to a chiral intermediate for the coibacin D synthesis, but the achieved enantioselectivity was low (58% ee) [102].

5.2 *Other Bioactive Substances*

Allylation of 2′,4′-difluorobiphenyl-4-carbaldehyde with allyltrichlorosilane catalyzed by (*S*,*R*)-**50** was exploited in the synthesis of the flobufen metabolite (Scheme 17) [114]. Optionally, allylation of benzaldehyde under the same conditions could be also used for a synthesis of dapoxetine [114]. Analogously, allylation of 2-thiophenecarbaldehyde with allyltrichlorosilane catalyzed by (*S*,*R*)-**50** was the basis for a new synthesis of duloxetine (Scheme 18) [115].

6 Conclusion

There is no doubt that achiral and chiral compounds possessing the pyridine *N*-oxide moiety could participate as catalysts in a wide range of organocatalyzed asymmetric reactions under rather mild reaction conditions. A rather large versatility of these catalysts in promoting both traditional reactions and, especially, new asymmetric reactions

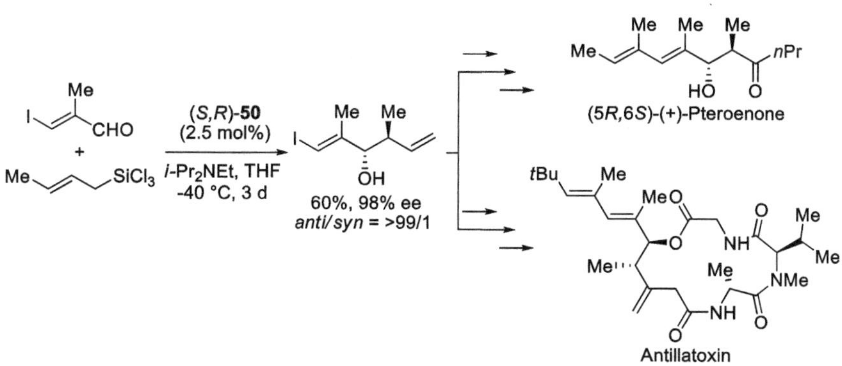

Scheme 16 Synthesis of (+)-pteroenone and antillatoxin

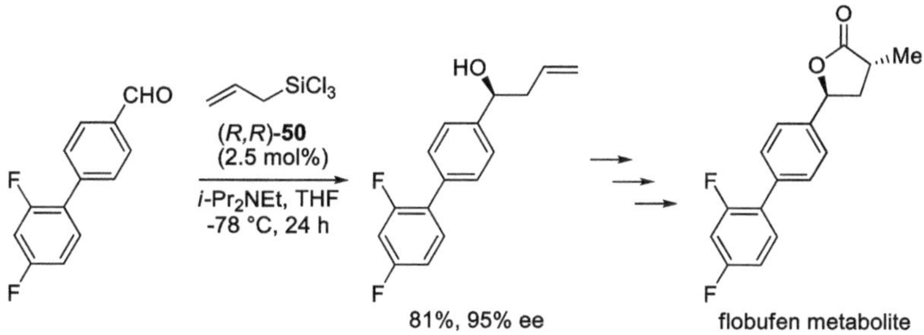

Scheme 17 Synthesis of the flobufen metabolite

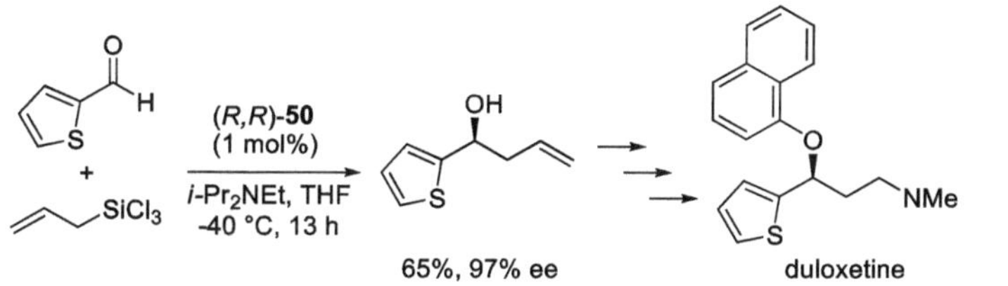

Scheme 18 Synthesis of duloxetine

is rewarding. The practical benefits of these reactions, especially in the area of asymmetric synthesis, include excellent enantioselectivity and activity providing useful chiral synthons with high enantiopurity, the use of cheap and readily available materials, mild reaction temperatures, operational simplicity, and last but not least, applicability in syntheses of natural or pharmaceutically active substances. There is no doubt that the future of the *N*-oxide family of catalysts, primarily in asymmetric transformations, is promising and will become one of the standard chemical tools in the field of organic synthesis.

References

1. Dalko PI, Moisa L (2001) Angew Chem Int Ed Engl 40:3726
2. Jha SC, Joshi NN (2002) ARKIVOC:167
3. Chelucci V, Thummel RP (2002) Chem Rev 102:3129
4. Malkov AV, Kočovský P (2003) Curr Org Chem 7:1737
5. Denmark SE, Fu J (2003) Chem Rev 103:2763
6. Dalko PI, Moisa L (2004) Angew Chem Int Ed Engl 43:5138
7. Kočovský P, Malkov AV (2004) Russ Chem Bull Int Ed 53:1806
8. Chelucci V, Murineddu G, Pinna GA (2004) Tetrahedron Asymmetry 15:1373
9. Rendler S, Oesterreich M (2005) Synlett:1727
10. Orito Y, Nakajima M (2006) Synlett:1391
11. Malkov AV, Kočovský P (2007) Eur J Org Chem:29
12. Liu X, Lin L, Feng X (2013) Acc Chem Res 44:574
13. Sutar RL, Joshi NN (2013) Tetrahedron Asymmetry 24:1345
14. Antkowiac WZ, Gessner WP (1979) Tetrahedron Lett 20:1931
15. Herrmann A, Hedman H, Rosén J, Jansson D, Haraldsson B, Hellenäs K-E (2012) J Nat Prod 75:1690
16. O'Donnell G, Poeschl R, Zimhony O, Gunaratnam M, Moreira JBO, Neidle S, Evangenopoulos D, Bhakta S, Malkinson JP, Boshoff HI, Lenaerts A, Gibbonds S (2009) J Nat Prod 72:360
17. Palani A, Shapiro S, Josien H, Bara T, Clader JW, Greenlee WJ, Cox K, Strizki JM, Baroudy BM (2002) J Med Chem 45:3143
18. Nantermet PG, Burgez CS, Robinson K, Pellicore JM, Newton CL, Deng JZ, Selnick HG, Lewis SD, Lucas BJ, Krueger JA, Miller-Stein C, White RB, Wong B, McMasters DR, Wallace AA, Lynch Jr JJ, Yan Y, Chen Z, Kuo L, Gardell SJ, Shafer JA, Vacca JP, Lyle TA (2005) Bioorg Med Chem Lett 15:2771
19. Mfuh AM, Larionov OV (2015) Curr Med Chem 22:2819
20. Balzarini J, Stevens M, De Clercq E, Schols D, Pannecouque C (2005) J Antimicrob Chemother 55:135
21. Brycki B, Szafran M (1984) J Chem Soc Perkin 2:223
22. Chmurzyński L (1992) J Solution Chem 21:171
23. Chmurzyński L, Liwo A, Barcyński P (1996) Anal Chim Acta 335:147
24. Chmurzyński L, Wawryniak G, Warnke Z (1997) J Heterocycl Chem 34:215
25. Ducháčková L, Kadlčíková A, Kotora M, Roithová J (2010) J Am Chem Soc 132:12660
26. Abraham MH, Honcharova L, Rocco SA, Acree Jr WE, De Fina KM (2011) New J Chem 35:930
27. Bergbreiter DE, Ortiz-Acosta D (2008) Tetrahedron Lett 49:5608
28. Hagiwara H, Inoguchi H, Fukushima M, Hoshi T, Suzuki T (2005) Synlett:2388
29. Denmark SE, Fan Y (2005) J Org Chem 70:9667
30. Murray JI, Woscholski R, Spivey AC (2014) Chem Commun 50:13608
31. Yoshida K, Takao K (2014) Tetrahedron Lett 55:6861
32. Yoshida K, Fujino Y, Itatsu Y, Inoue H, Kanoko Y, Takao K (2016) Tetrahedron Lett 57:627
33. Ando M, Emoto S (1975) Bull Soc Chim Jpn 48:1655
34. Ryzhkov AV, Rodina LL (2006) Russ J Gen Chem 76:126
35. Belousova IA, Simanenko YS, Savelova VA (2001) Russ J Org Chem 37:969
36. Belousova IA, Savelova VA, Simanenko YS, Panchenko BV (2002) Russ J Org Chem 38:111
37. Shiina I, Ushiyama H, Yamada Y, Kawakita Y, Nakata K (2008) Chem Asian J 3:454
38. Fujii M, Honda A (1992) J Heterocycl Chem 29:931
39. Fujii M, Honda A (1992) Chem Express 7:329
40. Nakajima M, Saito M, Shiro M, Hashimoto S (1998) J Am Chem Soc 120:6419
41. Shimada T, Kina A, Ikeda S, Hayashi T (2002) Org Lett 4:2799
42. Kina A, Shimada T, Hayashi T (2004) Adv Synth Catal 346:1169
43. Kwak J, Ohk J, Jung Y, Chang S (2012) J Am Chem Soc 134:17778

44. Jiao Z, Feng X, Liu B, Chen F, Zhang G, Jiang Y (2003) Eur J Org Chem:3818
45. Malkov AV, Dufková L, Farrugia L, Kočovský P (2003) Angew Chem Int Ed Engl 42:3674
46. Hrdina R, Stará GI, Dufková L, Mitchell S, Císařová I, Kotora M (2006) Tetrahedron 62:96
47. Hrdina R, Kadlčíková A, Valterová I, Hodačová J, Kotora M (2006) Tetrahedron Asymmetry 17:3185
48. Hrdina R, Valterová I, Hodačová J, Císařová I, Kotora M (2007) Adv Synth Catal 349:822
49. Hrdina R, Dračínský M, Valterová I, Hodačová J, Císařová I, Kotora M (2008) Adv Synth Catal 350:1449
50. Kadlčíková A, Hrdina R, Valterová I, Kotora M (2009) Adv Synth Catal 351:1279
51. Bai B, Shen L, Ren J, Zhu J-H (2012) Adv Synth Catal 354:354
52. Deng Y, Pan W, Pei N-Y, Li L-J, Bai B, Zhu J-H (2013) Tetrahedron 69:10431
53. Pignataro L, Benaglia M, Cinquini M, Cozzi F, Celentano G (2005) Chirality 17:396
54. Nakajima M, Yokota T, Saito M, Hashimoto S (2004) Tetrahedron Lett 45:61
55. Gao D-W, Gu Q, You S-L (2016) ACS Catal 4:2741
56. Malkov AV, Orsini M, Pernazza D, Muir WK, Langer V, Meghani P, Kočovský P (2002) Org Lett 4:1047
57. Malkov AV, Bell M, Orsini M, Pernazza D, Massa A, Herrmann P, Meghani P, Kočovský P (2003) J Org Chem 68:9659
58. Malkov AV, Bell M, Vassieu M, Bugatti V, Kočovský P (2003) J Mol Catal A Chem 196:179
59. Malkov AV, Bell M, Casteluzzo F, Kočovský P (2005) Org Lett 7:3219
60. Chelucci G, Belmonte N, Benaglia M, Pignataro L (2007) Tetrahedron Lett 48:4037
61. Denmark SE, Fan Y, Eastgate MD (2005) J Org Chem 70:5235
62. Diana MB, Marchetti M, Melloni G (1995) Tetrahedron Asymmetry 6:1175
63. Wong W-L, Lee C-S, Leung H-K, Kwong H-L (2004) Org Biomol Chem 2:1967
64. Pignataro L, Benaglia M, Annunziata R, Cinquini M, Cozzi F (2006) J Org Chem 71:1458
65. Derdau V, Laschat S, Hupe E, König WA, Dix I, Jones PG (1999) Eur J Inorg Chem:1001
66. Denmark SE, Fan Y (2002) J Am Chem Soc 124:4233
67. Bolm C, Ewald M, Zehnder M, Neuburger MA (1992) Chem Ber 125:453
68. Bolm C, Ewald M, Felder M, Schlingloff G (1992) Chem Ber 125:1169
69. Boyd RD, Sharma DN, Sbircea L, Murphy D, Malone FJ, James LS, Allen CRC, Hamilton TGJ (2010) Org Biomol Chem 8:1081
70. Gnanamani E, Someshwar N, Ramanathan RC (2012) Adv Synth Catal 354:2101
71. Gnanamani E, Someshwar N, Sanjeevi J, Ramanathan RC (2014) Adv Synth Catal 356:2219
72. Neniškis A, Stončius S (2015) Eur J Org Chem:6359
73. Gaecía-Flores F, Flores-Michel LS, Juaristi E (2006) Tetrahedron Lett 47:8235
74. Ikai T, Moro M, Maeda K, Kanoh S (2001) React Funct Polym 71:1055
75. Tao B, Lo MM-C, Fu GC (2001) J Am Chem Soc 123:353
76. Chai Q, Song C, Sun Z, Ma Y, Ma C, Dai Y, Andrus BM (2006) Tetrahedron Lett 47:8611
77. Fulton JR, Glover JE, Kamara L, Rowlands GJ (2011) Chem Commun 47:433
78. Takenaka N, Sarangthem RS, Captain B (2008) Angew Chem Int Ed Engl 47:9708
79. Chen J, Takenaka N (2009) Chem A Eur J 15:7268
80. Chen J, Captain B, Takenaka N (2011) Org Lett 13:2011
81. O'Hora SP, Incerti-Pradillos CA, Kabeshov AM, Shipilovskikh AS, Rubtsov EA, Elsegood RJM, Malkov VA (2015) Chem A Eur J 21:4551
82. Denmark SE, Fan Y (2006) Tetrahedron Asymmetry 17:687
83. Malkov AV, Ramírez-López P, Biedermannová L, Rulíšek L, Dufková L, Kotora M, Zhu F, Kočovský P (2008) J Am Chem Soc 130:5341
84. Cadart T, Koukal P, Kotora M (2014) Eur J Org Chem:7556
85. Hessler F, Betík R, Kadlčíková A, Belle R, Kotora M (2014) Eur J Org Chem:7245
86. Shimada T, Kina A, Hayashi T (2003) J Org Chem 68:6329
87. Malkov VA, Barlóg M, Miller-Potucká L, Kabeshov MA, Farrugia LJ, Kočovský P (2012) Chem A Eur J 18:6873

88. Malkov AV, Ramírez-Lopéz P, Biedermannová L, Rulíšek L, Dufková L, Kotora M, Zhu F, Kočovský P (2008) J Am Chem Soc 130:5431
89. Bai B, Zhu H-J, Pan W (2012) Tetrahedron 68:6829
90. Malkov AV, Kabeshov AM, Barlog M, Kočovský P (2009) Chem A Eur J 15:1570
91. Malkov AV, Stončius S, Bell M, Castelluzo F, Ramírez-Lopéz P, Biedermannová L, Langer V, Rulíšek L, Kočovský P (2013) Chem A Eur J 19:9167
92. Chelucci G, Baldino S, Pinna GA, Benaglia M, Buffa L, Guizzetti S (2008) Tetrahedron 64:7574
93. Incerti-Pradillos CA, Kabeshov AM, O'Hora SP, Shipilovskikh AS, Rubtsov EA, Drobkova VA, Baladina SY, Malkov AV (2016) Chem A Eur J 21:14390
94. Kadlčíková A, Vlašaná K, Ducháčková L, Roithová J, Kotora M (2010) Chem A Eur J 16:9442
95. Vlašaná K, Hrdina R, Valterová I, Kotora M (2010) Eur J Org Chem:7040
96. Malkov AV, Barzog M, Jewkes Y, Mikušek J, Kočovský P (2011) J Org Chem 76:4800
97. Koukal P, Ulč J, Nečas D, Kotora M (2016) Eur J Org Chem:2110
98. Matoušová E, Koukal P, Formánek B, Kotora M (2016) Org Lett 18:5656
99. Koukal P, Kotora M (2015) Chem A Eur J 21:7408
100. Vlašaná K, Betík R, Valterová I, Nečas D, Kotora M (2016) Curr Organocatal 3:301
101. Hrdina R, Boyd T, Valterová I, Hodačová J, Kotora M (2008) Synlett:3141
102. Kolská K, Ghavre M, Pour M, Hybelbauerová S, Kotora M (2016) Asian J Org Chem 5:646
103. Nakajima M, Saito M, Hashimoto S (2002) Tetrahedron Asymmetry 13:2449
104. Peng Y, Takenaka N (2012) Chem Rec 13:28
105. Lu T, Zhu R, An Y, Wheeler SE (2012) J Am Chem Soc 134:3095
106. Sugiura M, Sato N, Sonoda Y, Kotani S, Nakajima M (2010) Chem Asian J 5:478
107. Kadlčíková A, Vlašaná K, Kotora M (2011) Collect Czech Chem Commun 76:415
108. Chen F-X, Qin B, Feng X, Zhang G, Jiang Y (2004) Tetrahedron 60:10449
109. Jiao Z, Feng X, Liu B, Chen F, Zhang G, Jiang Y (2003) Eur J Org Chem:3818
110. Su Z, Hu C, Feng X (2006) Tetrahedron 62:3818
111. Sugiura M, Kumahara M, Nakajima M (2009) Chem Commun:3585
112. Diana MB, Marchetti M, Melloni G (1995) Tetrahedron Asymmetry 6:1175
113. Matoušová E, Koukal P, Formánek B, Kotora M (2016) Org Lett 18:5656
114. Hessler F, Korotvička A, Nečas D, Valterová I, Kotora M (2014) Eur J Org Chem:2543
115. Motloch P, Valterová I, Kotora M (2014) Adv Synth Catal 356:199

Top Heterocycl Chem (2017) 53: 59–84
DOI: 10.1007/7081_2017_1

Published online: 13 April 2017

Transition Metal-Catalyzed C–H Functionalization of Heterocyclic *N*-Oxides

David E. Stephens and Oleg V. Larionov

Abstract Transition metal-catalyzed C–H-functionalization reactions of heterocyclic *N*-oxides are reviewed. Arylation, alkenylation, alkylation, cyanation, amidation, and sulfonylation reactions of azine, azole, and non-aromatic *N*-oxides are discussed with an emphasis on the regioselectivity of the C2–H, as well as distal C–H functionalization, of heterocyclic *N*-oxides. The review primarily focuses on the advances from the past decade.

Keywords Alkenylation • Alkylation • Alkynylation • Allylation • Amination • Arylation • Azine • azole • Azole *N*-oxide • Catalysis • C–H functionalization • Cyanation • Heterocycles • Iodination • *N*-oxide • Olefination • Palladium • Pyridine • quinoline • Selenylation • Sulfonylation • Sulfoximination • Transition metal

Contents

D.E. Stephens and O.V. Larionov (✉)
The University of Texas at San Antonio, San Antonio, TX, USA
e-mail: oleg.larionov@utsa.edu

1 Introduction

The development of new methods of functionalization of *N*-heterocycles can lead to improved syntheses of pharmaceuticals and agrochemicals [1–3], advanced materials, and ligands for transition metal catalysis [4, 5]. Methods for selective C–H functionalization of heterocyclic *N*-oxides have attracted significant attention in the past 10–15 years. Due to the inherent electronic bias, heterocyclic *N*-oxides tend to react primarily at the C2 position, and many C2-selective C–H-functionalization reactions of *N*-oxides have recently been described. In contrast, the regioselectivity of C–H functionalization of unoxidized *N*-heterocycles can be more difficult to control. The regioselectivity of C–H functionalization of heterocyclic *N*-oxides, along with the ease of preparation [6–8], and the versatility [9–15] of the *N*-oxide functionality make them a prominent class of *N*-heterocyclic synthetic intermediates.

This chapter will first discuss C–H-functionalization reactions of azole *N*-oxides, and C2–H functionalization of azine *N*-oxides. Distal C–H functionalization of azine *N*-oxides will follow. The discussion will primarily be limited to reactions that proceed without the concomitant deoxygenation of the *N*-oxide moiety. The focus of the review will be on the synthetic utility and applications of the reactions. Several other recent reviews have discussed azine C2–H functionalization [16–23], as well as the distal C–H functionalization [24–26] of azines and azoles. Within each section, examples will be listed in chronological manner, unless more than one report has been combined to simplify the presentation. Several common heterocyclic *N*-oxide motifs that are discussed in this review are shown in Scheme 1.

2 Azole *N*-Oxides

In 2008, Fagnou and coworkers described the stepwise regioselective arylation of thiazole *N*-oxides [27]. Previously, the arylation of thiazoles had been known to proceed with low regioselectivity. Due to their inherent C–H acidity at the C2 position and the π-nucleophilicity at C5, bis-arylation, or C5 selective arylation, had only been reported. Fagnou and coworkers proposed that the π-nucleophilicity could be responsible for the C2/C5 selectivity as shown in Scheme 2. Upon

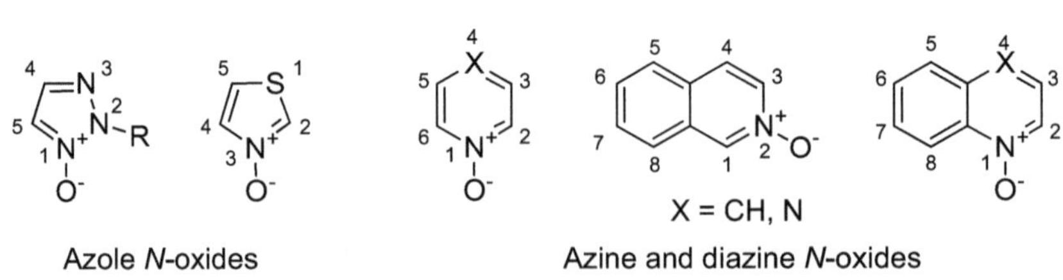

Scheme 1 Common heterocyclic *N*-oxide motifs

Scheme 2 HOMO contributions and Pd-catalyzed arylation of thiazole *N*-oxides

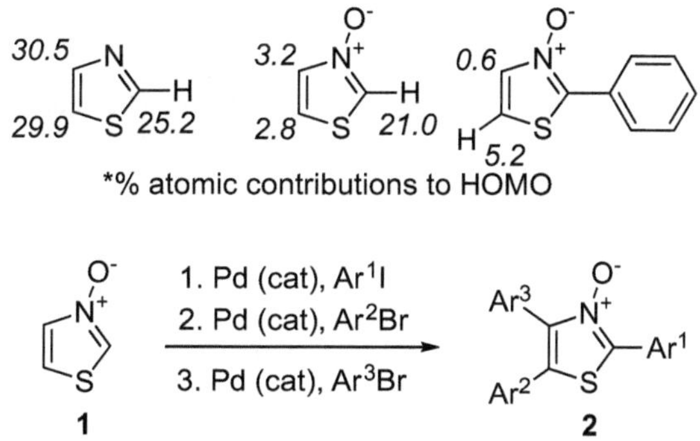

Scheme 3 Arylation of protected imidazolone *N*-oxides

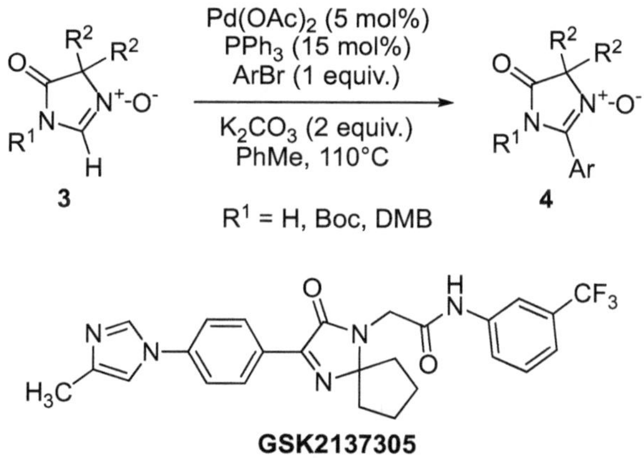

oxidation of the nitrogen, a reliable C2 > C5 > C4 order of regioselectivity was observed. This order of selectivity also remained unchanged when the C2-position was substituted. Using this method sequential functionalization of thiazole *N*-oxide (**1**) can be achieved, leading to a highly diversified thiazole scaffold. Fourteen highly substituted thiazole *N*-oxides **2** were produced (Scheme 2).

Cyclic nitrone *N*-oxides are also attractive substrates for C–H arylation, due to their versatility as precursors for α-amino acids. In 2012, Zhao, Wang, and coworkers reported the first arylation of imidazolone *N*-oxides **3** with aryl bromides (Scheme 3) [28]. Using palladium acetate/triphenylphosphine as a catalyst, the authors were able to produce the arylated *N*-protected imidazolone derivatives **4** in up to 88% yield. With the unprotected nitrogen of the amido group, the yield was reduced to 49%. The reaction performs best with aryl bromides, with lower yields observed for aryl iodides, chlorides, and triflates. Twelve 2-arylimidazolne *N*-oxides were produced in 64–95% yields. The authors then used this methodology to synthesize the glycine transporter type-1 (GlyT1) inhibitor GSK2137305.

Chavant, Blandin, and coworkers reported the arylation of cyclic nitrones **5** (Scheme 4) [29]. The authors optimized the reaction in anisole utilizing a Pd^0

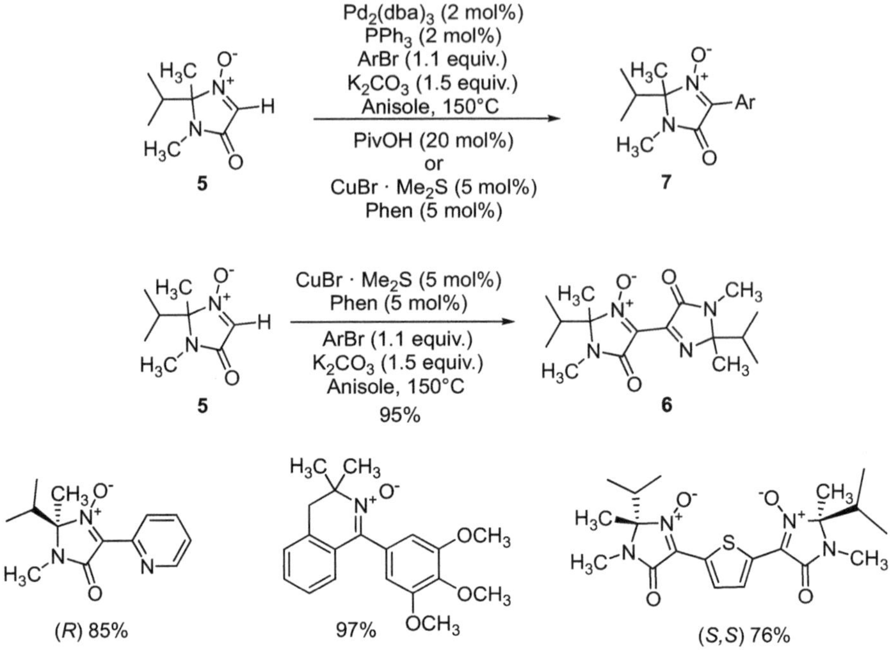

Scheme 4 Direct arylation and homocoupling of cyclic nitrones

catalyst at 150°C with either pivalic acid or a CuBr·Me_2S/1,10-phenanthroline (Phen) system. Interestingly, in the absence of $Pd_2(dba)_3$ and PPh_3, only the product of C5 homocoupling **6** was obtained in 95% yield. The system tolerates a wide range of aryl bromides as coupling partners. The C2 stereocenter was not affected, when enantiomerically pure cyclic nitrones **5** were used as substrates. Arylation products of type **7** were synthesized in 68–99% yields.

Kuang and coworkers reported the C5-selective olefination and arylation of 1,2,3-triazole *N*-oxides **8** (Scheme 5) [30]. The authors began their optimization using Fujiwara–Moritani conditions [31] as a starting point. Silver carbonate proved to be the oxidant of choice, with pyridine as an additive in a 20% *t*-BuOH/1,4-dioxane mixture. In addition to high C5 regioselectivity, the reaction exhibits high *E*-diastereoselectivity with styrenes and acrylates as coupling partners, as only *E*-isomers **9** were formed as products.

The authors also observed formation of small amounts of arylation products **10** when performing the solvent screen. Further optimization allowed for an efficient arylation with *p*-xylene, *m*-xylene, benzene, and 1,2-dichlorobenzene. Notably, no reaction took place with an unoxidized 1,2,3-triazole, highlighting the crucial role of the *N*-oxide functionality.

In a follow-up study, Kuang and coworkers reported the dual C–H/C–H cross coupling of 2-aryl-1,2,3-triazole *N*-oxides **8** and azine *N*-oxides en route to bis-*N*-oxides **11** (Scheme 6) [32]. The scope of the azine *N*-oxides included isoquinoline

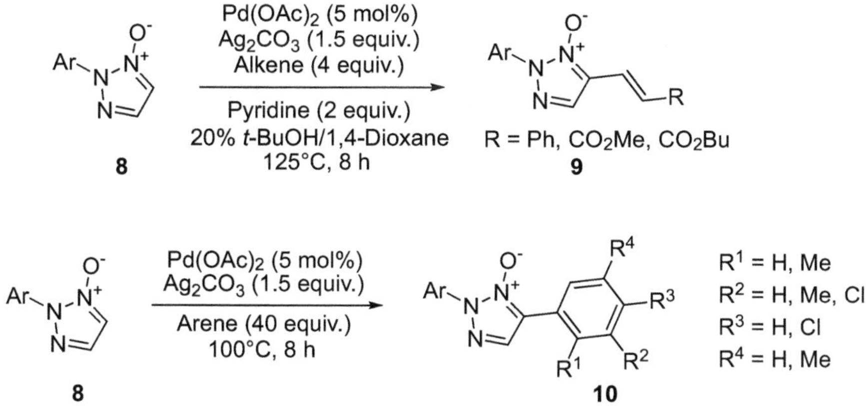

Scheme 5 Olefination and arylation of 2-substituted 1,2,3-triazole *N*-oxides

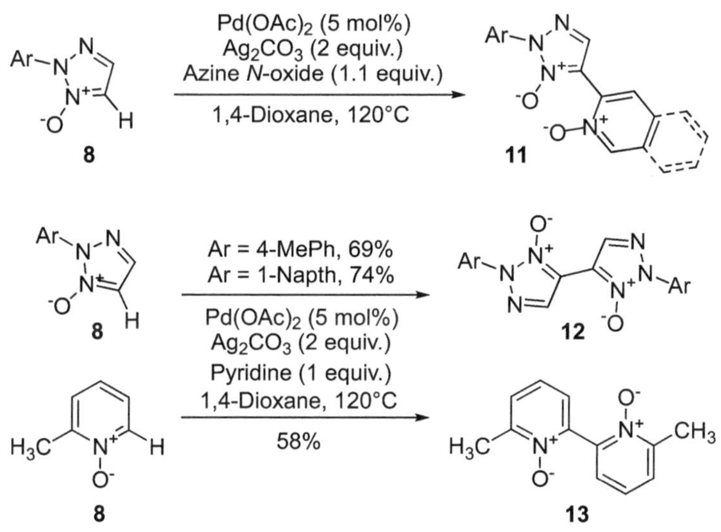

Scheme 6 Pd-catalyzed oxidative C–H/C–H homocoupling of 1,2,3-triazole *N*-oxides and pyridine *N*-oxides

N-oxide and substituted pyridine *N*-oxides. Interestingly, no reaction was observed with quinoline *N*-oxides or 2-substitutued pyridine *N*-oxides. While pyridine *N*-oxides reacted exclusively at C2 position, isoquinoline *N*-oxide reacted at C3 position. This regioselectivity is unusual for isoquinoline *N*-oxide, since most of the C–H-functionalization reactions occur at C1 position. A number of functional groups are well tolerated, including chloro, fluoro, methyl, and alkoxy groups. The homocoupling reactions of 2-aryl-1,2,3-triazole *N*-oxides and pyridine *N*-oxides were also observed in the reaction mixture. Upon optimization of the homocoupling reaction, it was found that addition of 1 equiv. of pyridine doubled the overall yield of products **12** and **13** in 48 h.

Kuang and coworkers were able to extend the methodology to the cross coupling of azoles and azine *N*-oxides (Scheme 7) [33]. The authors were able to utilize *N*1-substituted 1,2,3-triazoles and other 5-memebered heterocycles for the coupling with azine *N*-oxides. An interesting dichotomy was reported with isoquinoline *N*-oxide and 1,2,3-triazoles or furans/thiophenes. With 1,2,3-triazoles, the C3-substituted isoquinoline derivative was exclusively obtained, whereas furans or thiophenes yielded only the C1-substituted isoquinoline *N*-oxide product. C2-substituted pyridine *N*-oxides were suitable substrates; quinoline *N*-oxides, on the other hand, did not react. Deuterium labeling experiments showed that a fast H/D exchange took place in the azole, while no H/D exchange was observed in the pyridine *N*-oxide.

This result was interpreted as an indication of a fast and reversible palladation of the azole that is then followed by a slow reaction with pyridine *N*-oxide.

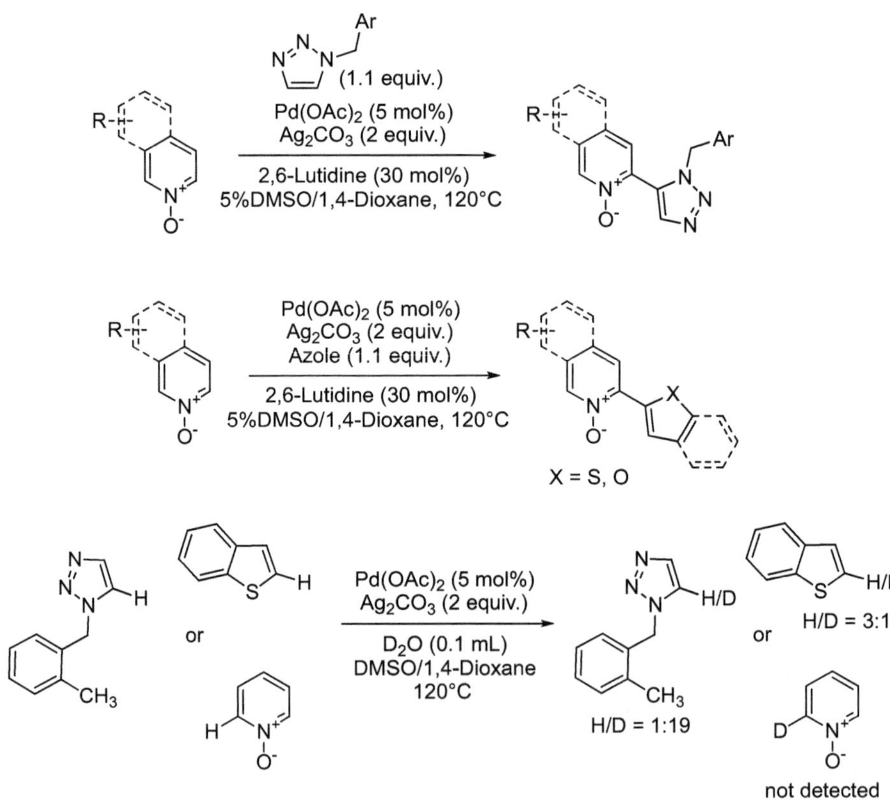

Scheme 7 Coupling of azine *N*-oxides with azoles and H/D exchange

3 Azine *N*-Oxides

3.1 Introduction of Heteroatom Substituents

Li and coworkers described the dehydrogenative cross coupling of azine *N*-oxides and cyclic amines (Scheme 8) [34]. It was noted during optimization studies that either 10 mol% $Cu(OAc)_2$ with 2 equiv. of Ag_2CO_3 or the same reagents in the opposite amounts gave satisfactory yields. The authors were able to extend the scope to the *N*-heteroarylated five-, six-, and seven-membered lactams **14**, as well as *N*-heteroarylated cyclic amines **15**. A double C–N coupling with piperazine gave rise to product **16** in 72% yield. Substituted quinoline, quinoxaline, and pyridine *N*-oxides were shown to be suitable coupling partners. Nitro, chloro, bromo, methoxy, methyl, and phenyl substituents were well tolerated. The reaction of 3-methylquinoline *N*-oxide with caprolactam produced the desired cross-coupling product in 85% yield, indicating that sterically hindered products can be prepared using this method. The low value of the kinetic isotope effect ($k_H/k_D = 1.3$) indicated that the C–H bond cleavage was not the rate-limiting step. A Cu^I/Cu^{III} catalytic cycle was proposed.

In 2013, Cui, Wu, and coworkers reported a C–H sulfonylation of quinoline *N*-oxides (Scheme 9) [35]. The reaction is enabled by CuI as a catalyst in the presence

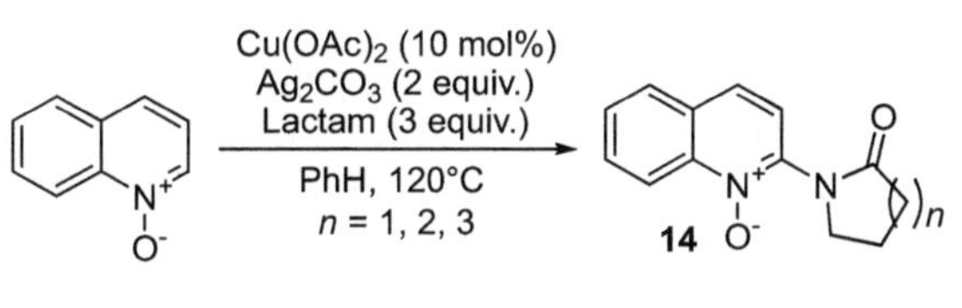

Catalyst (10 mol%)	Oxidant (2 equiv.)	Yield %
none	$Cu(OAc)_2$	trace
none	Ag_2CO_3	trace
$Cu(OAc)_2$	Ag_2CO_3	93
Ag_2CO_3	$Cu(OAc)_2$	81

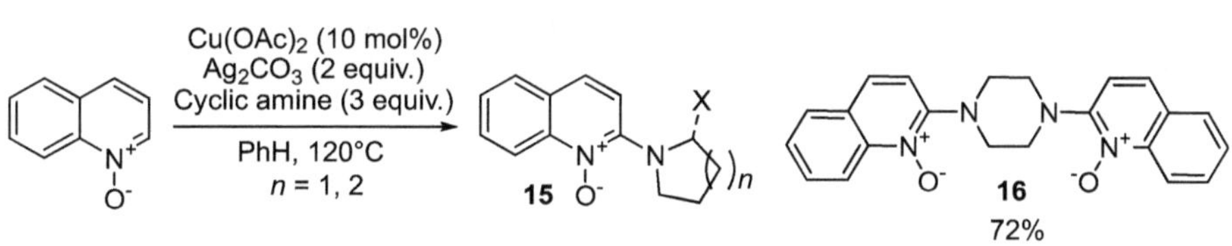

Scheme 8 Dehydrogenative amidation and optimization conditions

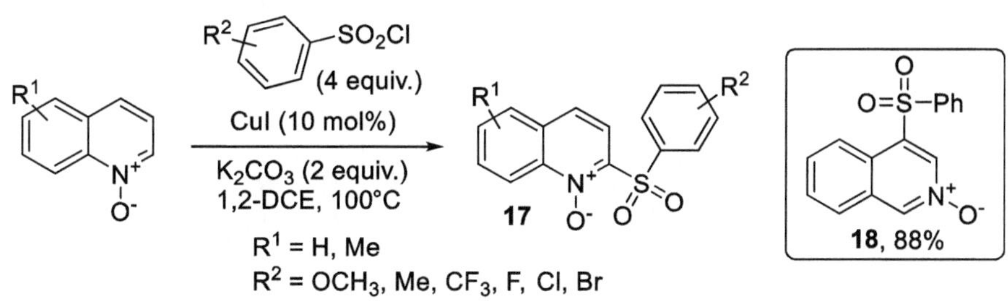

Scheme 9 C–H sulfonylation of azine *N*-oxides with *p*-tolylsulfonyl chloride

of potassium carbonate. The scope includes quinoline *N*-oxide, as well as 4- and 6-methylquinoline *N*-oxides, and isoquinoline *N*-oxide. The method is not compatible with *N*-oxides derived from pyrazine, pyrimidine, 1-methyl imidazole, pyridine, and their derivatives. Arenesulfonyl chlorides with trifluoromethyl, chloro, methoxy, fluoro, and methyl groups in the *para*- and *meta*-positions proved to be suitable sulfonylation reagents. Interestingly, although C2-sulfonylated products of type **17** are formed with quinoline *N*-oxides, only C4–H-sulfonylation product **18** is observed with isoquinoline *N*-oxide in 88% yield.

Cui and Wei further disclosed a copper-catalyzed amination of azine *N*-oxides with secondary amines (Scheme 10) [36]. Air was used as the oxidant in this protocol. When run under an atmosphere of nitrogen, only a 16% yield was obtained at 50°C. Both cyclic and acyclic amines were suitable coupling partners. Quinoline *N*-oxides bearing chloro, bromo, nitro, methoxy, and methyl substituents in the 3, 4, or 6 positions were converted to the C2-amination products **19** in good yields. 8-Methylquinoline *N*-oxide and 2-phenylpyridine *N*-oxide, on the other hand, provided trace amounts of the expected amination products. In addition, the *N*-oxides derived from isoquinoline, quinoxaline, and 1,10-phenanthroline gave the C2-amination products in good yields. Two notable exceptions were also reported. Thus, in the first case, when 4-bromoquionline *N*-oxide was subjected to the amination, the C2/C4 diamine product **20** was obtained in 31% yield. In the second case, the reaction with 4-nitroquinoline afforded the product of the double amination at C2 and C8 positions **21** in 80% yield.

In 2015, Li and coworkers reported the copper-catalyzed electrophilic amination of azine *N*-oxides (Scheme 11) [37]. The amination is effected by *O*-benzoyl hydroxylamines with $Cu(OAc)_2$ and Ag_2CO_3 as catalysts in *t*-BuOH at 80°C. The scope of the reaction includes the *N*-oxides of quinoline, quinoxaline, and isoquinoline.

Bolm and coworkers recently reported the copper-catalyzed C2–H amination of azine *N*-oxides using sulfoximines as coupling partners (Scheme 12) [38].

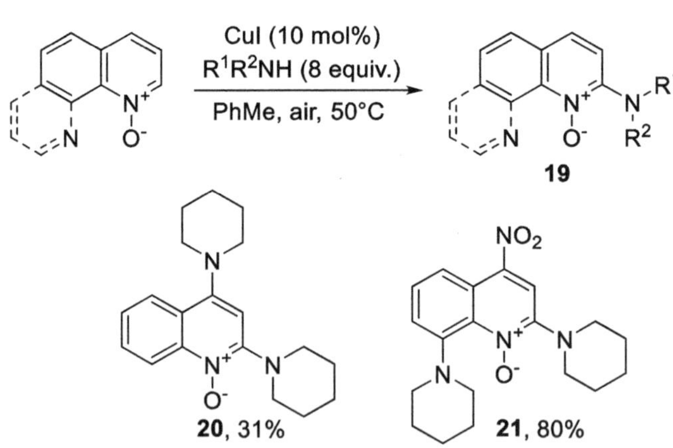

Scheme 10 Copper-catalyzed amination of azine *N*-oxides

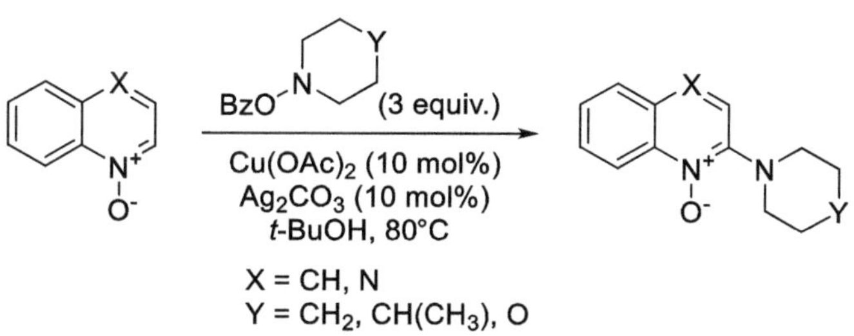

Scheme 11 Electrophilic amination of azine *N*-oxides with *O*-benzylhydroxylamine

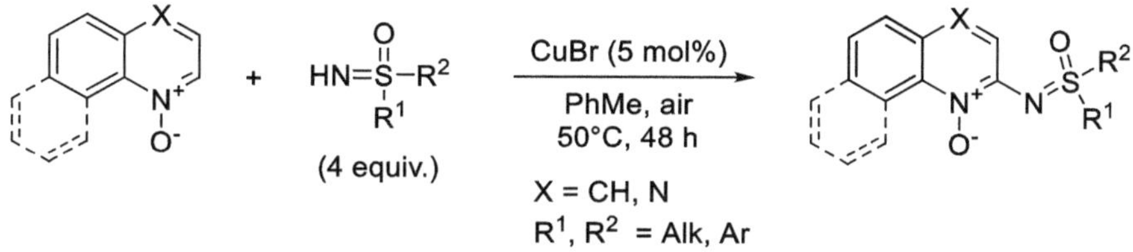

Scheme 12 Sulfoximination of azine *N*-oxides

Copper(I) bromide was used as a catalyst with air as the terminal oxidant at 50°C. *N*-Oxides of quinoline and quinoxaline were found to be suitable substrates, while only trace product was observed with pyridine *N*-oxide. The reaction exhibited a broad scope with respect to the substituted sulfoximines.

3.2 *C–C-Bond Formation via Cross Coupling with Heterocyclic* N*-Oxides*

In 2005, Fagnou and coworkers provided the first example of an efficient transition metal-catalyzed C2–H functionalization of substituted pyridine *N*-oxides with a range of aryl bromides (Scheme 13) [39–41]. The method required excess pyridine *N*-oxide (4 equiv.). The catalytic system consisted of 5 mol% $Pd(OAc)_2$ and $Pt\text{-}Bu_3$ (15 mol%). A number of substituted aryl bromides were shown to be suitable substrates. In addition to pyridine *N*-oxide, 4-nitro and 4-methoxypyridine *N*-oxides could also be utilized as coupling partners.

In order to determine if the mechanism followed an electrophilic aromatic substitution (S_EAr) pathway, a competition experiment between 4-methoxy- and 4-nitropyridine *N*-oxide was performed (Scheme 13). Since formation of product **22** was greatly favored over product **23**, a S_EAr mechanism was ruled out.

Fagnou and Leclerc subsequently reported the arylation of diazine *N*-oxides, with high regioselectivity, as well as the bis-arylation of pyrazine *N*-oxide (Scheme 14) [42]. In all cases the aryl group was introduced at the carbon adjacent to the N–

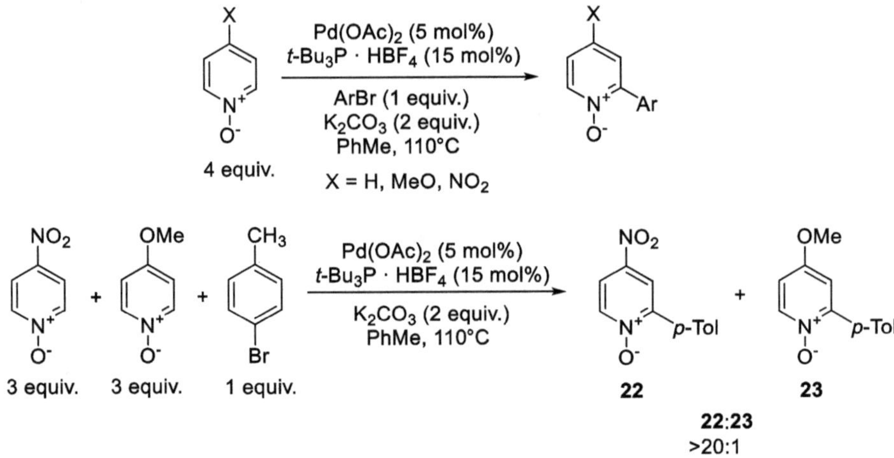

Scheme 13 Pd-catalyzed C2–H arylation of pyridine *N*-oxides

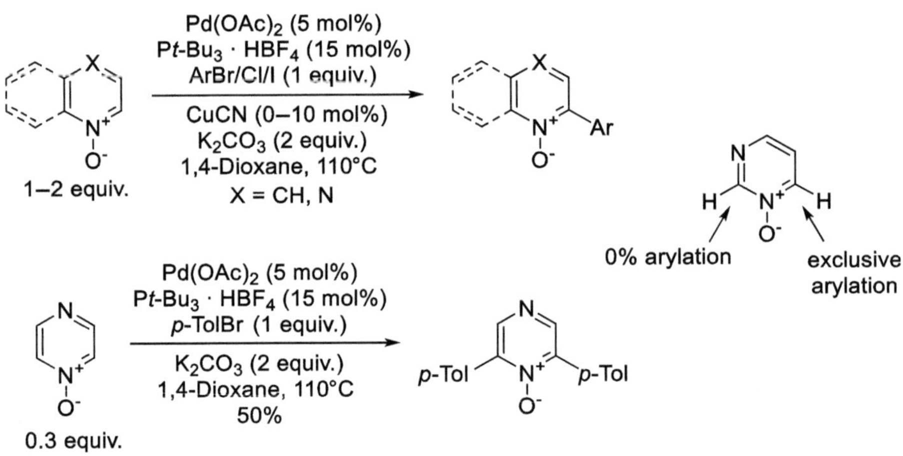

Scheme 14 Pd-catalyzed arylation of diazine *N*-oxides

O moiety but not between the two nitrogens in pyrimidine *N*-oxide. Aryl bromides, chlorides, and iodides were suitable coupling partners. Addition of 10 mol% CuBr or CuCN increased the yield in the case of pyrimidine *N*-oxide C–H arylation that suffered from catalyst inhibition. With aryl iodides, Ag_2CO_3 (0.5 equiv.) had to be used as an additive. The scope of the *N*-heterocyclic substrates included the *N*-oxides of pyridazine, pyrimidine, pyrazine, and quinoxaline. Fagnou and coworkers subsequently further extended the scope to include aryl triflates as coupling partners [43].

As an application of this methodology, Fagnou and coworkers reported the synthesis of two aporphine alkaloid analogues **24** and **25** (Scheme 15) [44].

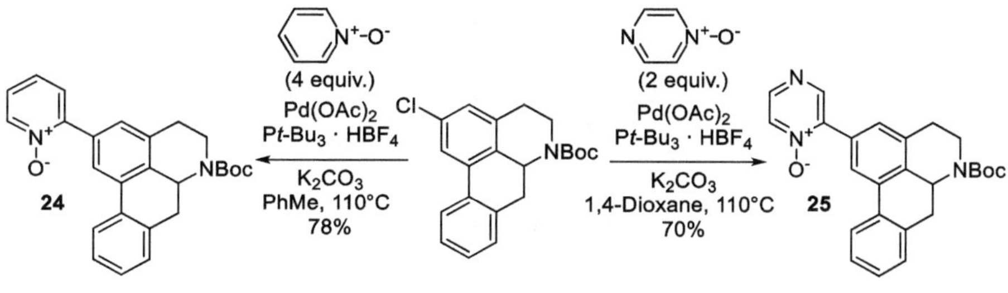

Scheme 15 Synthesis of aporphine alkaloid analogues **24** and **25**

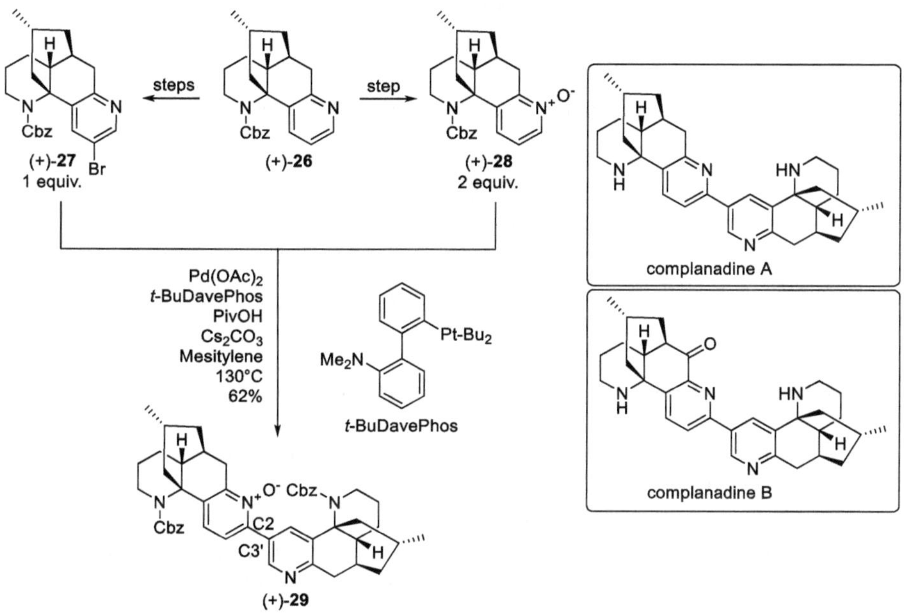

Scheme 16 C2–H functionalization of azine *N*-oxides in the synthesis of complanadines A and B

An interesting example of the utilization of azine *N*-oxide heteroarylation in natural product synthesis was presented by Tsukano in the synthesis of complanadines A and B (Scheme 16) [45]. The complanadines exhibit an unsymmetrical C2–C3′ bipyridyl linkage. In order to construct this linkage, the common intermediate (+)-**26** was converted to bromide (+)-**27** and *N*-oxide (+)-**28**. Pd-catalyzed cross coupling of intermediated (+)-**27** and (+)-**28** yielded an advanced intermediate (+)-**29** that was subsequently converted to complanadine A and complanadine B in one and two steps, respectively.

Daugulis and coworkers reported a CuI-catalyzed C–H arylation of pyridine *N*-oxides (Scheme 17) [46, 47]. While lithium *tert*-butoxide was required for iodoarenes, the reaction of pyridine *N*-oxide with 2-iodopyridine took place with

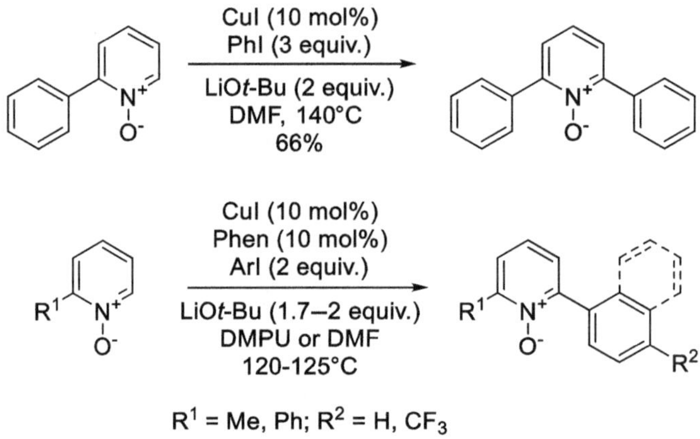

Scheme 17 CuI promoted arylation of 2-substituted pyridine *N*-oxide

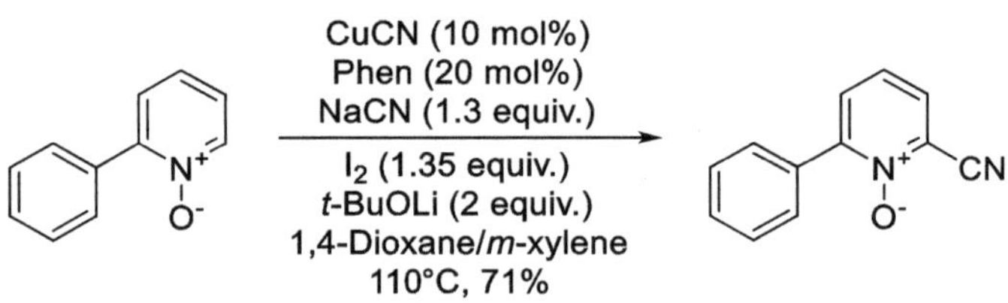

Scheme 18 Cu-catalyzed C–H cyanation of 2-phenylpyridine *N*-oxide

the weaker base K_3PO_4. You and coworkers also reported a copper-catalyzed C–H arylation of pyridine and quinoline *N*-oxides with *p*-bromotoluene [48].

In 2010, Daugulis and Do reported a single example of Cu-catalyzed cyanation of 2-phenylpyridine *N*-oxide (Scheme 18) [49]. The reaction was effected by iodine and sodium cyanide, with lithium *tert*-butoxide acting as a base.

Introduction of secondary alkyl groups via C–H functionalization of *N*-heterocycles remains a challenge. Ohmiya and Sawamura reported the Cu-catalyzed *E*-stereoselective allylic alkylation of electron-deficient arenes with secondary allylic phosphates (Scheme 19) [50]. 2-Phenylpyridine *N*-oxide was successfully alkylated using this method.

Hiyama, Nakao, and Kanyiva provided an example of the nickel-catalyzed *E*-selective C–H alkenylation of azine *N*-oxides (Scheme 20) [51]. The scope was limited to internal alkynes, while terminal alkynes did not participate in the reaction. C2 alkenylation was observed with pyridine *N*-oxides. Isoquinoline *N*-oxide, on the other hand, underwent alkenylation at C1 position. The *E*/*Z* selectivity was in the range of 13:1 to 99:1.

Chang and coworkers disclosed an *E*-selective alkenylation and a dual C–H/C–H arylation of azine *N*-oxides [52]. The reaction calls for use of excess (4 equiv.) *N*-oxide. $Pd(OAc)_2$ (10 mol%) was used as a catalyst, with 1.5 equiv. of Ag_2CO_3. The *N*-oxides of pyridine, pyridazine, pyrazine, and quinoxaline were suitable

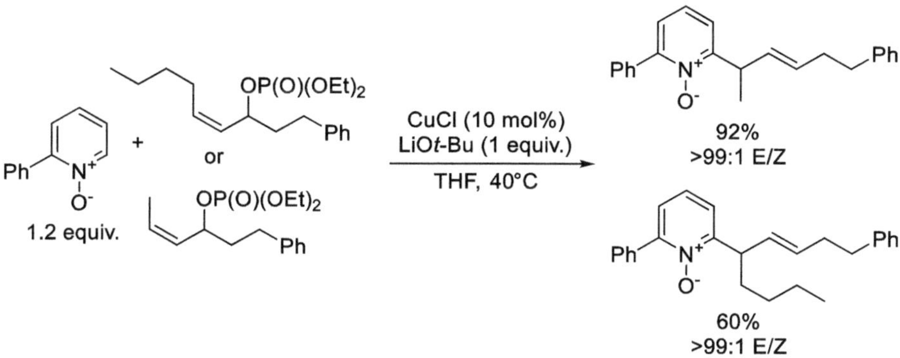

Scheme 19 Cu-catalyzed C2–H allylation of 2-phenylpyridine *N*-oxide

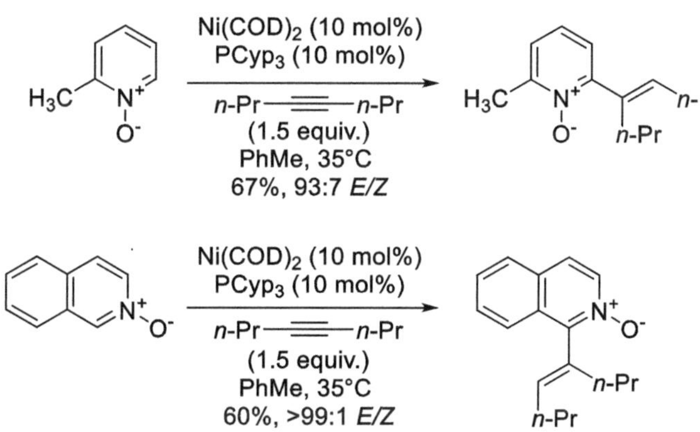

Scheme 20 Ni-catalyzed alkenylation of azine *N*-oxides

substrates. The scope of alkenes included terminal monosubstituted alkenes bearing electron-withdrawing groups, as well as styrene and *tert*-butylethylene (Scheme 21).

During the optimization of the C–H-alkenylation reaction, the authors noticed that a considerable amount of 2-phenylpyridine *N*-oxide was produced in benzene at 130°C. High *ortho*-selectivity is reported for the arylation of *N*-oxides derived from pyridines, quinoline, pyrazine, quinoxaline, and isoquinoline. The scope of the carbocyclic aromatic coupling partners included benzene, *o*- and *m*-xylene, *o*-difluorobenzene, and *o*-dichlorobenzene. The regioselectivity of the C–H functionalization in the substituted arenes was governed by the steric influence of the substituents (Scheme 21). The ratio of bis- to mono-arylation products formed from pyridine *N*-oxides ranged from 1:2.5 to 1:20. An X-ray structure of Pd complex **30** was also obtained (Scheme 22). Since complex **30** did not undergo the C2–H alkenylation, a conclusion was made that it is a resting species outside of the catalytic cycle.

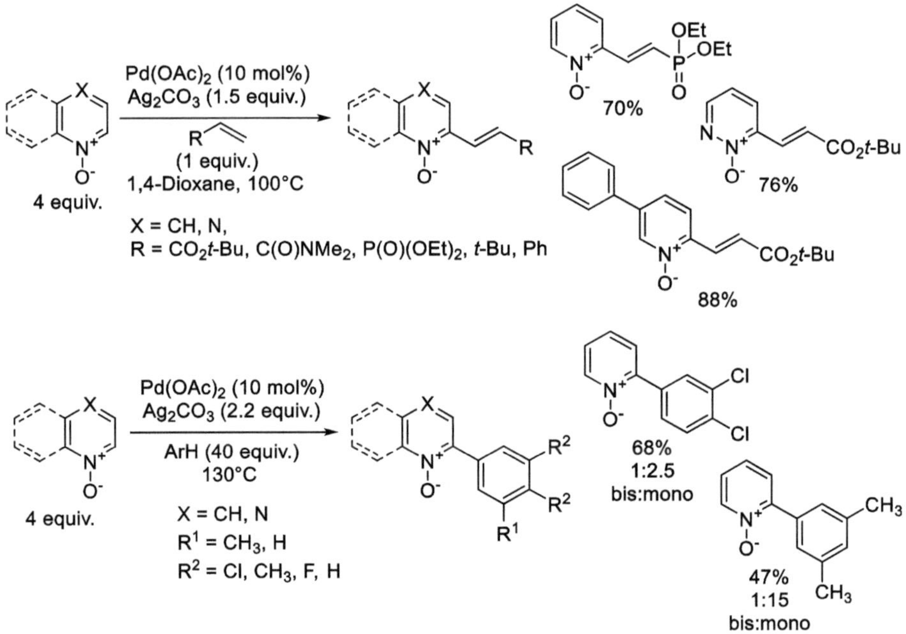

Scheme 21 Pd-catalyzed olefination and arylation of azine and diazine *N*-oxides

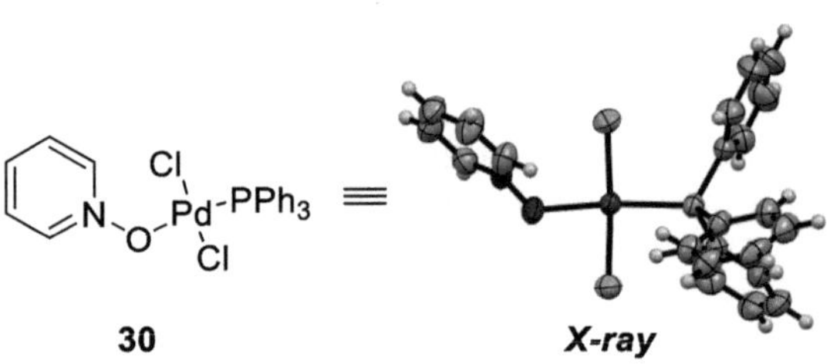

Scheme 22 Crystal structure of Pd(II)/*N*-oxide complex **30**

Due to the facile arylation of the indole ring at C2/C3 in azaindoles (Scheme 23), C–H functionalization of the pyridine ring represents a significant challenge. In 2009 Fagnou and Huestis addressed this regioselectivity problem by employing the *N*-oxide as a directing group for the coupling of aryl bromides and azaindoles (Scheme 23) [53]. The scope of the *N*-heterocyclic coupling partners included *N*-oxides derived from the *N*1-methyl-protected 6- and 7-azaindoles. Palladium acetate with DavePhos was found to be the catalytic system of choice for this reaction. Excess *N*-oxide (2 equiv.) was required to achieve good yields. A 22% increase in the yield was observed with the addition of 30 mol% pivalic acid. All arylation reactions were carried out with bromoarenes.

In 2010 Hu, You, and coworkers reported a palladium-catalyzed dual C–H/C–H oxidative cross-coupling reaction of quinoline, pyridine, and 2-methylquinoline *N*-

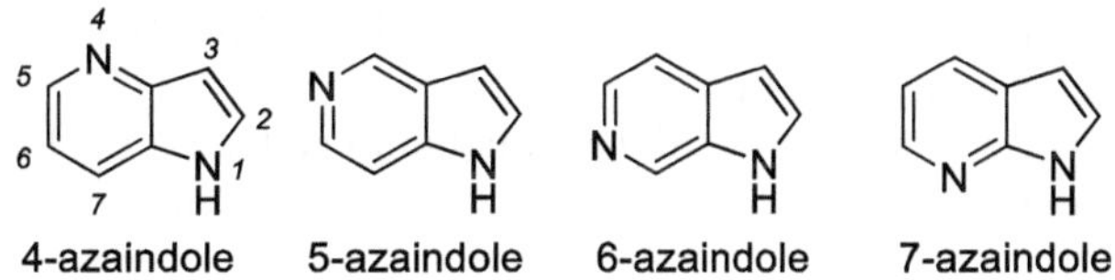

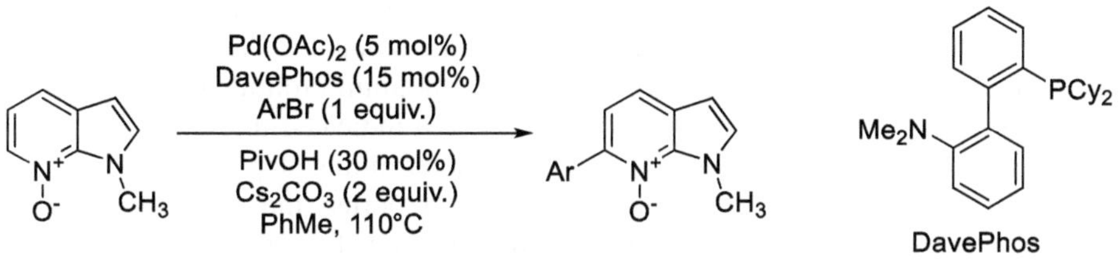

Scheme 23 Pd-catalyzed arylation of azaindole *N*-oxides

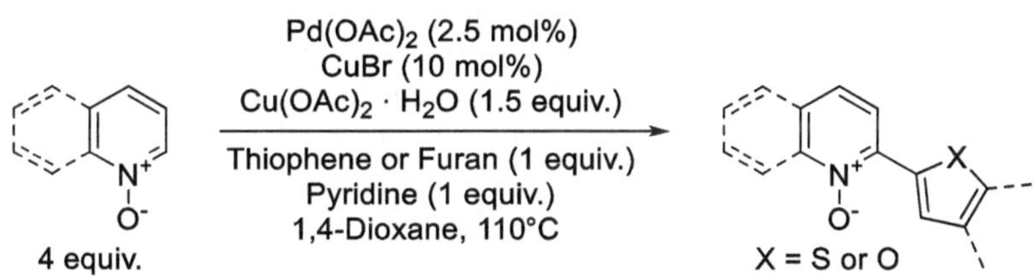

Scheme 24 Pd-catalyzed oxidative C–H/C–H cross-coupling of azine *N*-oxides with thiophenes and furans

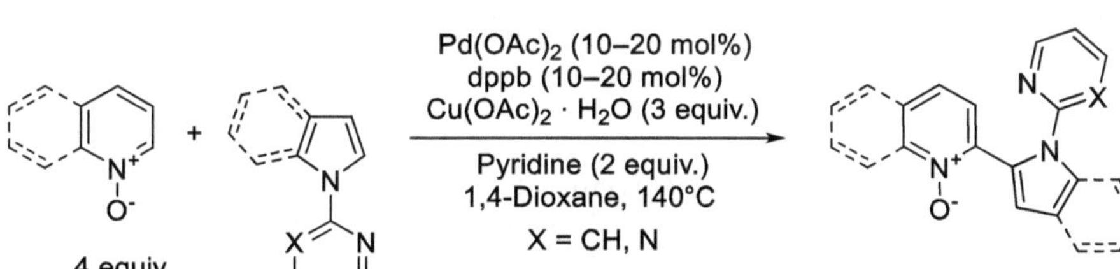

Scheme 25 Pd-catalyzed C–H/C–H cross-coupling reaction of azine *N*-oxides with azoles

oxides with substituted thiophenes and furans, as well as benzothiophene (Scheme 24) [54]. Excess (4 equiv.) azine *N*-oxide was necessary to effect the cross-coupling reaction with 35–80% yields. Only C2–H coupling was observed for both heterocycles. The reaction required 10 mol% CuBr as a co-catalyst, in addition to Cu $(OAc)_2$ (1.5 equiv.) as an oxidant. The coupling of 2-methylthiophene and quinoline *N*-oxide was performed on a 2 g scale, highlighting the scalability of the reaction.

You further extended the methodology to pyrrole and indoles (Scheme 25) [55]. The 2-pyridyl or 2-pyrimidyl directing group was appended to the azole nitrogen to ensure C2 regioselectivity within the azole ring. As with thiophenes and furans, 4 equiv. of azine *N*-oxide (quinoline, quinoxaline, and pyridine *N*-

oxides) was required to ensure good yields. 1,4-bis(diphenylphosphino)butane (dppb) was used as a ligand in this case.

Li, Zhang, and coworkers reported the palladium-catalyzed coupling of pyridine *N*-oxides with indole that proceeds with C3 regioselectivity in the indole ring (Scheme 26) [56].

The authors began their optimization with the reaction conditions reported by You and Hu (Scheme 23) [54], which yielded the cross-coupled product in 12% yield. Further optimization resulted in changing the oxidant to Ag_2CO_3, increasing the amount of pyridine, and adding tetrabutylammonium bromide (TBAB). Large intermolecular kinetic isotope effect ($k_H/k_D = 3.6$) indicated that the C2–H bond cleavage in the *N*-oxide is involved in the rate-determining step.

Similarly Itami, Yamaguchi, and coworkers reported the cross coupling of indoles and pyrroles with azine *N*-oxides that proceeded with C3-regioselectivity in the indole and pyrrole rings (Scheme 27) [57]. The reaction conditions included 10 mol% of $Pd(OAc)_2$ as a catalyst, 3 equiv. of AgOAc as an oxidant, and 1 equiv. of 2,6-lutidine. Azine *N*-oxides were used in excess (4 equiv.). The reaction was successfully applied to the synthesis of eudistomin U in five steps from commercially available β-carboline (Scheme 27). The same team subsequently also applied the C3-selective coupling in their synthesis of *rac*-dragmacidin D (Scheme 28) [58].

You and coworkers also reported a C3-selective (in the indole ring) coupling of indoles and azine/diazine *N*-oxides (Scheme 29) [59].

Cui, Wu, and coworkers reported the alkylation of azine *N*-oxides with ethers and thioethers (Scheme 30) [60]. The reaction was proposed to proceed through a Pd^0/Pd^{II} catalytic cycle. The substrate scope included quinoline, isoquinoline, and pyridine *N*-oxides. One example of coupling with ethanol was provided; all other

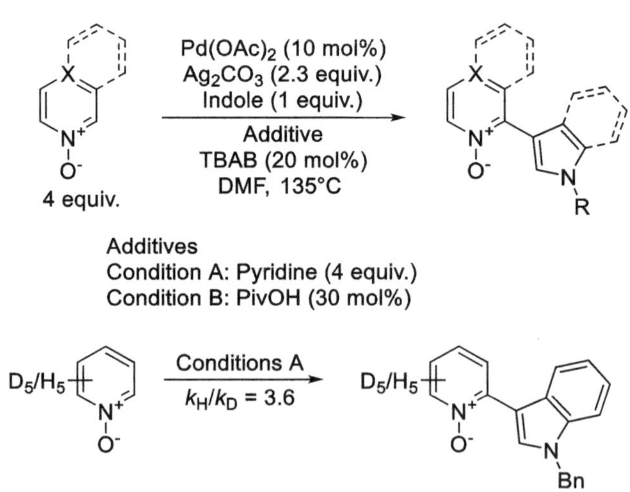

Scheme 26 Pd-catalyzed cross coupling of azine *N*-oxides with indoles

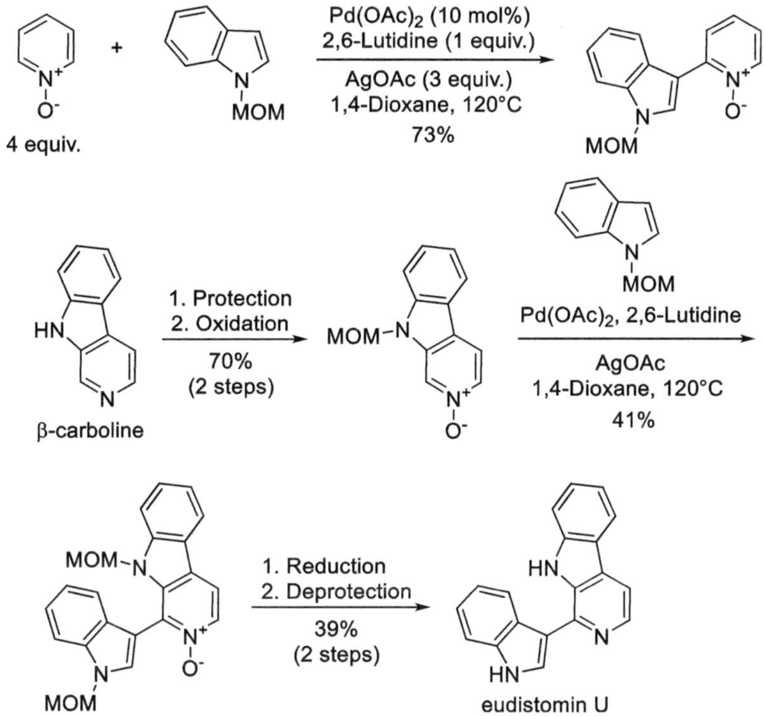

Scheme 27 Pd-catalyzed cross coupling of azine N-oxides with indoles and the synthesis of eudistomin U from β-carboline

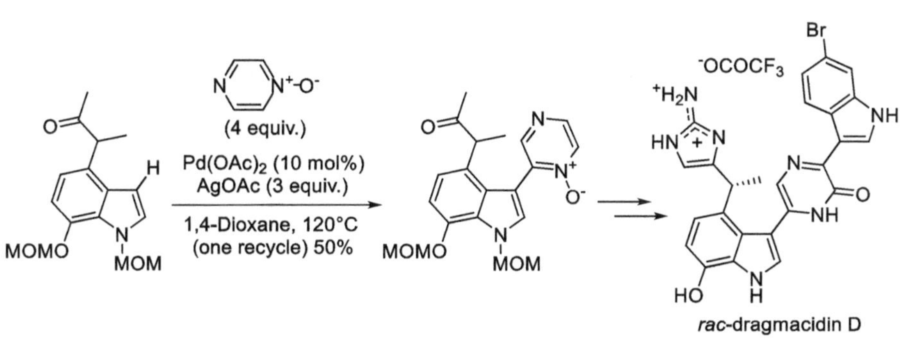

Scheme 28 C–H/C–H cross coupling en route to dragmacidin D

examples utilized cyclic ethers. The ethers and thioethers were used in a large excess as the solvent in this method.

In 2013, Fu and coworkers reported a Pd-catalyzed C–H alkylation of azine N-oxides with alkyl halides (Scheme 31) [61]. The initial optimization study focused on the reaction between 2-methylpyridine N-oxide and cyclohexyl bromide. Various palladium sources were tested, with $Pd(OAc)_2$dppf providing the best yield in

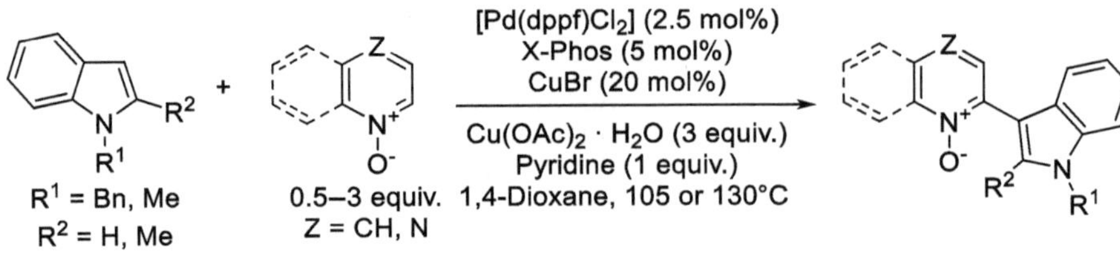

Scheme 29 Oxidative cross coupling of indoles and azine *N*-oxides

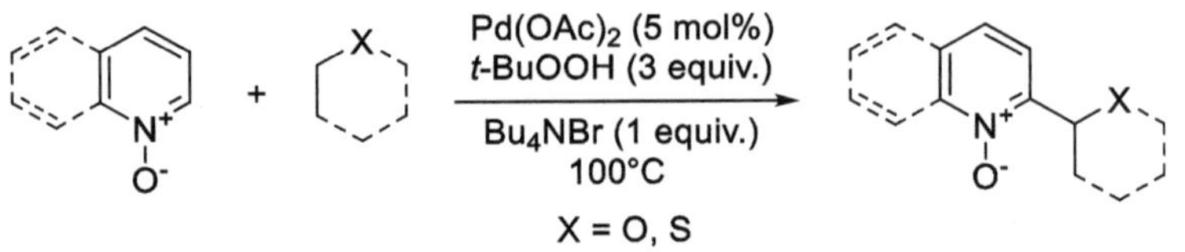

Scheme 30 Alkylation of azine *N*-oxides with ethers or ethanol

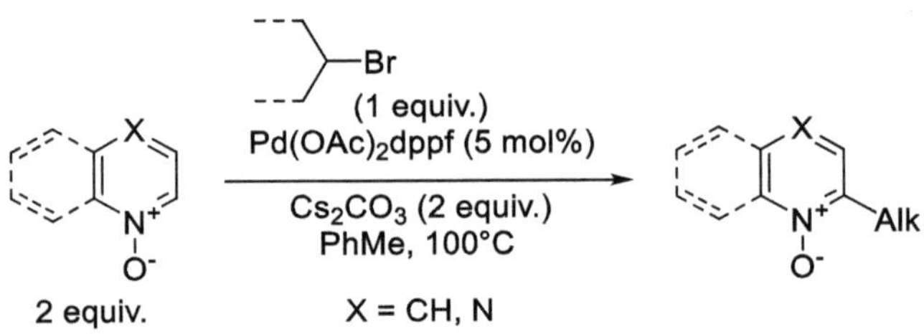

Scheme 31 Alkylation of azine *N*-oxides with secondary alkyl bromides

the presence of Cs_2CO_3 as a base at 100°C. The reaction scope included a number of cycloalkyl and alkyl bromides, as well aza- and oxacycloalkyl bromides. Mechanistic studies pointed to the involvement of a radical-type mechanism. Similarly, a Pd/dppf-catalyzed C2–H alkylation of azine *N*-oxides with cyclohexyl iodide was reported by Zhou and coworkers as a demonstration of the utility of their catalytic method of C–H alkylation of *N*-heterocycles [62].

In 2016, Jain and Jha reported the Cu-catalyzed *ortho*-alkylation of azine *N*-oxides with *N*-tosylhydrazones under microwave irradiation (MW) (Scheme 32) [63]. The reaction is carried out in the presence of lithium *tert*-butoxide as a base under microwave irradiation for 1 h at 100°C. The substrate scope included chloro, methyl, and phenyl-substituted pyridine *N*-oxides, as well as isoquinoline and quinoline *N*-oxides. Primary and secondary *N*-tosylhydrazones were suitable reagents.

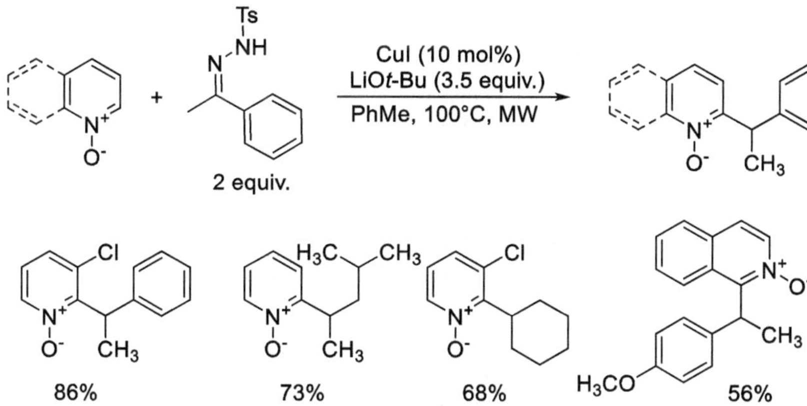

Scheme 32 Copper-catalyzed alkylation of azine *N*-oxides with *N*-tosylhydrazones

3.3 Distal C–H Functionalization of Azine N-Oxides

While in most applications the *N*-oxide moiety of *N*-heterocycles serves as an effective directing group for the C2–H functionalization, the moiety can also direct the incoming substituents to the C8 position of the quinoline scaffold. The first example of a catalytic C8–H functionalization of quinoline *N*-oxides was reported in 2014 by Chang and coworkers (Scheme 33) [64]. Chang and coworkers previously used a Rh dimer to catalyze C8–H arylation of quinolines [65]. When $[RhCp^*Cl_2]_2/AgNTf_2$ was used as a catalyst with *N*-iodosuccinimide (NIS) as an electrophilic iodine source, regioselective iodination at the C8 position of quinoline *N*-oxide took place. Several substituted quinoline *N*-oxides along with quinoxaline *N*-oxide were efficiently iodinated in C8 position. Interestingly, Chang and coworkers were also able to develop a C8-selective amidation of quinoline *N*-oxides with sulfonyl azides. In this case, $[Cp^*IrCl_2]_2/AgNTf_2$ was found to be the catalyst of choice (Scheme 33). Both aliphatic and aromatic sulfonyl azides were shown to be suitable amidation reagents.

More recently, Chang and coworkers developed the rhodium-catalyzed C8–H amidation of quinoline *N*-oxides using *N*-chlorocarbamates (Scheme 34) [66]. The required *N*-chlorocarbamates could be conveniently synthesized from the corresponding carbamates and trichloroisocyanuric acid (TCCA) and subjected to the C8–H-amidation reaction in a one-pot fashion on gram scale.

Later that year, Chang and coworkers published the direct alkylation and alkynylation of quinoline *N*-oxides in the C8 position utilizing hypervalent iodine reagent **31** and diazoesters **32**, respectively (Scheme 35) [67]. Unlike other reported C8–H alkynylations [68–71], in which C8 substitution is accompanied by reduction of the *N*-oxide moiety, the utilization of TIPS-EBX (**31**) retains the *N*-oxide (Scheme 35). The pregenerated cationic $[RhCp^*(MeCN)_3][SbF_6]_2$ was used as a catalyst. A number of diverse di- and tricyclic quinoline *N*-oxide analogues were

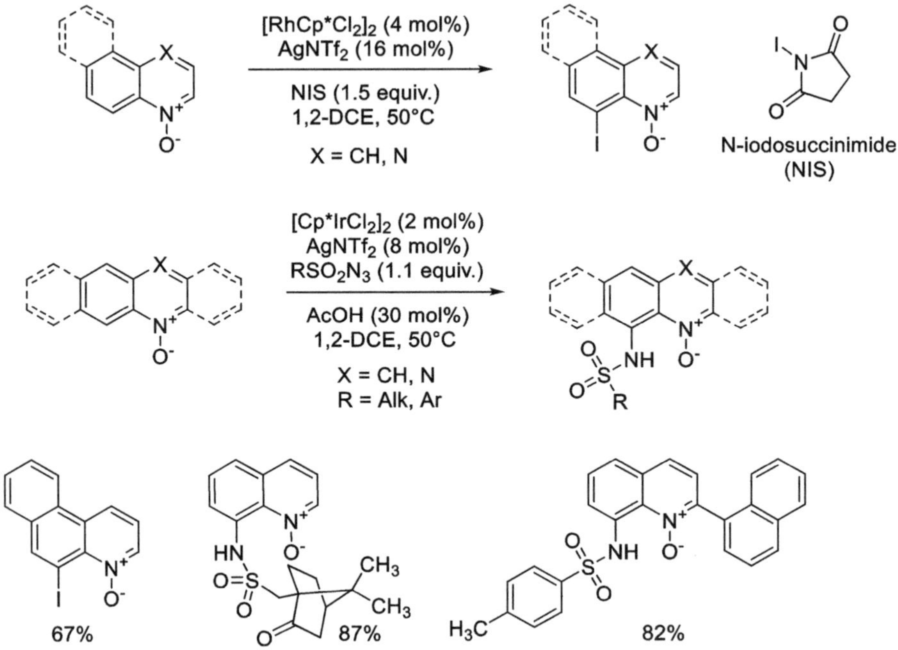

Scheme 33 C8–H iodination and amidation of azine *N*-oxides

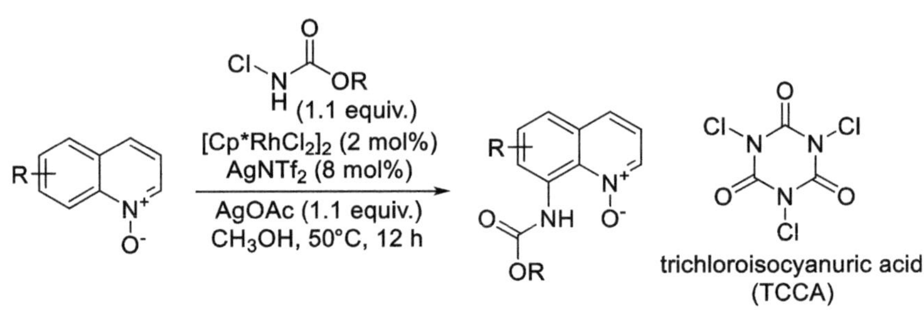

Scheme 34 Rh-catalyzed C8–H amidation of azine *N*-oxides with *N*-chlorocarbamates

efficiently alkynylated in C8 position. Similar conditions were also adapted for the C8 alkylation with diazoesters **32**. In this case, the catalyst was formed in situ. The functional group tolerance and the *N*-oxide substrate scope are similar for both reactions.

Shibata and Matsuo disclosed a C8-selective alkenylation of quinoline *N*-oxides with diarylacetylenes (Scheme 36) [72]. $[Rh(COD)_2]OTf$/xylylBINAP was used as a catalytic system at 135°C in chlorobenzene with symmetrical diarylacetylenes. The alkenyl moiety in C8 position of the products was produced with *E*/*Z* ratios in the range of 3:1 to >20:1.

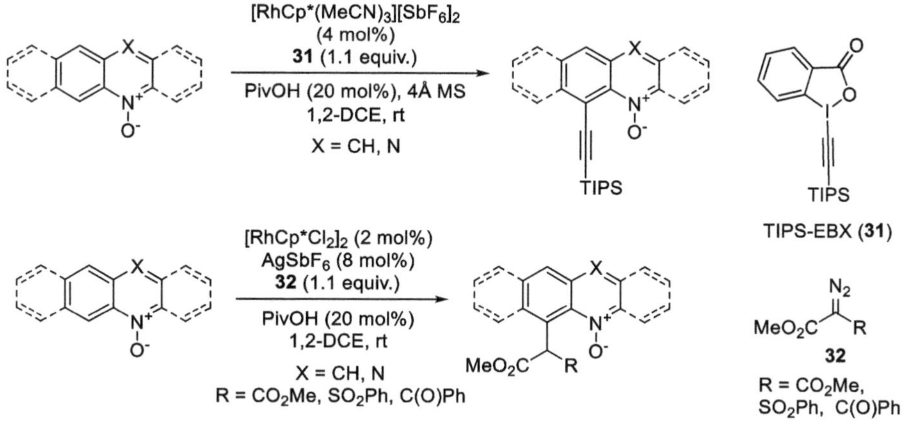

Scheme 35 C8–H alkynylation and alkylation of azine *N*-oxides

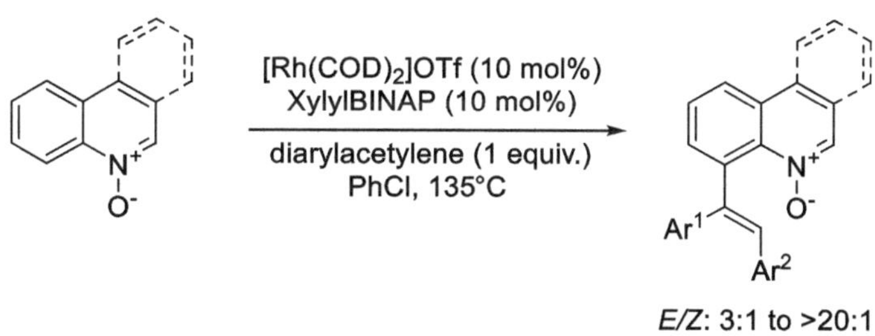

Scheme 36 C8–H alkenylation of azine *N*-oxides with diarylacetylenes

Larionov and coworkers developed a Pd-catalyzed C8–H arylation of quinoline *N*-oxides (Scheme 37) [73]. An interesting effect of the solvent on the regioselectivity of the arylation was first uncovered. When DMF or *t*-BuOH was utilized, C2 selectivity was preferred. However, when AcOH (as low as 10 equiv.) was used as a solvent, the C8–H arylation became predominant with the C8/C2 selectivity in excess of 20:1 with only 5 mol% $Pd(OAc)_2$. The authors also found that the addition of water (up to 40 equiv.) improved the overall yield by 10–20%. The reaction was carried out under thermal and microwave conditions. A two-step synthesis (oxidation followed by arylation) was performed in the microwave in under 2 h.

In 2015 Chang and coworkers published a protocol for the C–H arylation of benzamides, acrylamides, and quinoline *N*-oxides. The method called for use of $[IrCp * Cl_2]_2$ (5 mol%) as a catalyst in 2,2,2-trifluoroethanol with aryldiazonium salts as the arylating reagents [74]. C8-regioselelctivity was observed for the arylation of quinoline *N*-oxides (Scheme 38).

Larionov and coworkers observed that in the absence of an aryl iodide, C8–H-homocoupling products were obtained from quinoline *N*-oxides (Scheme 39) [75].

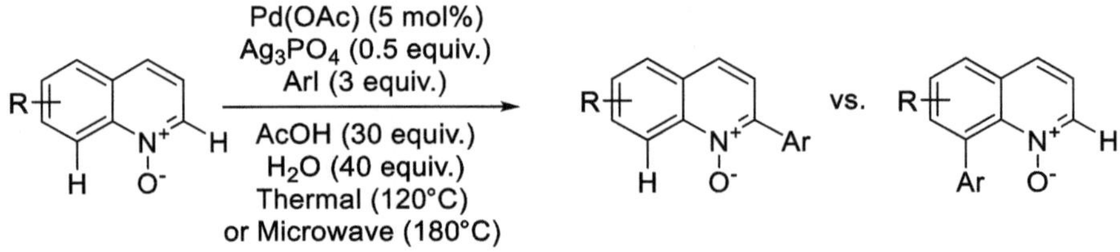

AgX	Solvent	SM	C_2	C_8
AgOAc (3)	DMF	29	36	5
AgOAc (3)	*t*-BuOH	50	41	7
Ag_2CO_3 (0.5)	AcOH	29	3	23
Ag_3PO_4 (0.5)	AcOH	8	<2	78
Ag_3PO_4 (0.5)	AcOH/H_2O	1	4	95

Scheme 37 Pd-catalyzed C8–H arylation of quinoline *N*-oxides

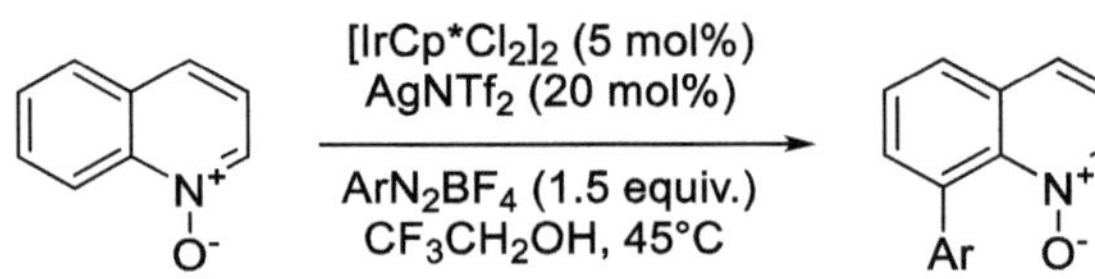

Scheme 38 Ir-catalyzed C8–H arylation of quinoline *N*-oxides

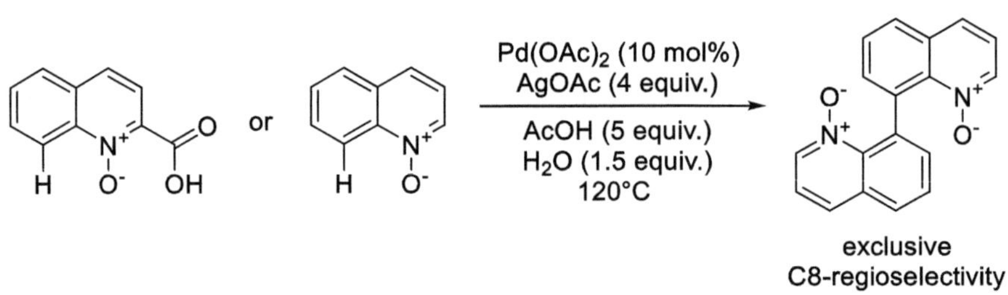

Scheme 39 Pd-catalyzed C8–H homocoupling of quinoline *N*-oxides

Silver acetate was used as an oxidant in the presence of AcOH (5 equiv.) and water (1.5 equiv.) at 120°C (Scheme 39). The reaction can be carried out on gram scale. Tandem C2 decarboxylation with C8–H homocoupling was also observed for 2-carboxyquinoline *N*-oxides. Mechanistic studies pointed to the reversible formation of a C8 palladacycle intermediate and suggested that the reductive elimination could be the turnover-limiting step. The method is complementary to the recently developed C2-selective dimerization of quinoline *N*-oxides [76, 77].

Li, Wan, and Yu recently developed a rhodium-catalyzed directed C–H selenylation of aromatic and heteroaromatic substrates. As a part of their synthetic scope study, the authors reported five examples of the C8–H selenation of quinoline

N-oxides with phenylselenyl chloride (Scheme 40) [78]. Exclusive selenation of quinoline *N*-oxides in C8 position was observed.

As an extension of their earlier publication on the C8–H acylation of quinoline *N*-oxides with the concomitant rearrangement to 2-quinolones [79], Chen, Cui, and Wu reported the Pd-catalyzed C8–H acylation of quinoline *N*-oxides with α-oxocarboxylic acids (Scheme 41) [80].

Inspired by the work by Muthusubramanian et al. on the silver-catalyzed decarboxylative acylation of azine *N*-oxides at C2 position [81], the authors began their study by optimizing the reaction of quinoline *N*-oxide and phenylglyoxylic acid. It was found that no inert atmosphere was necessary, while potassium persulfate was the oxidant of choice. Mechanistic studies suggested that palladacycle dimer **33** [79] may be formed in the reaction, while a noticeable kinetic isotope effect (k_H/k_D of 3.6) further suggested that the C8–H bond cleavage was the rate-limiting step.

As a follow-up to their work on the cobalt-catalyzed C8–H alkenylation with the concomitant oxygen transfer [71], Sundararaju and coworkers reported a chemodivergent C8–H functionalization of quinoline *N*-oxides (Scheme 42) [82]. The authors discovered that a reaction of quinoline *N*-oxides with allylic alcohols produced 8-allylquinoline *N*-oxides with a cobalt catalyst derived from $[Cp^*CoI_2]_2$. On the other hand, a rhodium catalyst derived from $[Cp * RhCl_2]_2$ directed the reaction to 8-(3-oxoalkyl)quinoline *N*-oxides. The unsubstituted allyl group was introduced in C8 position using allyl ethyl carbonate. 3-Arylallyl groups were installed using the corresponding 2-arylallyl alcohols. The cobalt catalyst was used in both cases. When $[Cp^*RhCl_2]_2$ was used with 2-arylallyl alcohols,

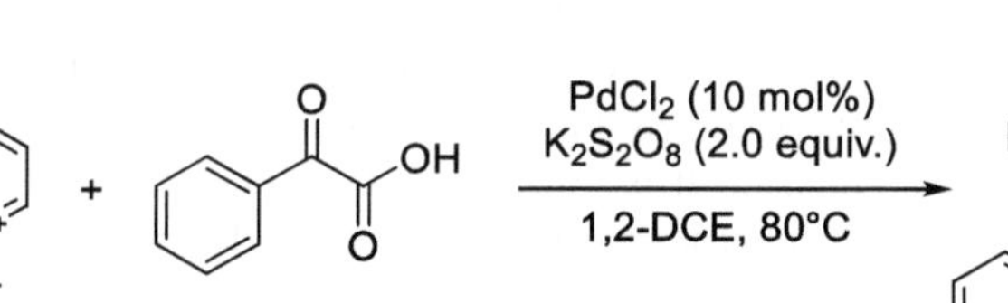

Scheme 40 C8–H selenylation of quinoline *N*-oxides

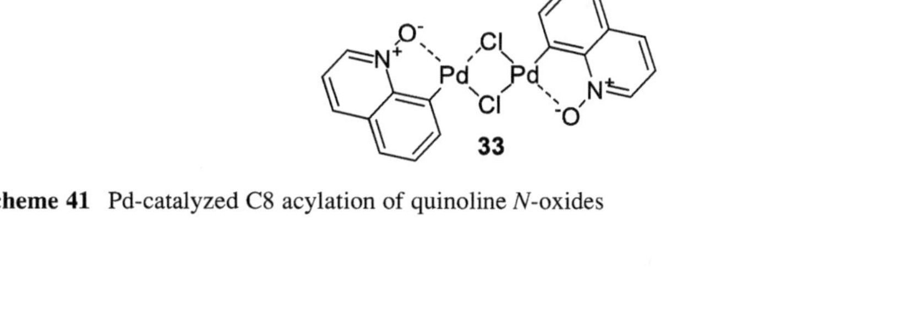

Scheme 41 Pd-catalyzed C8 acylation of quinoline *N*-oxides

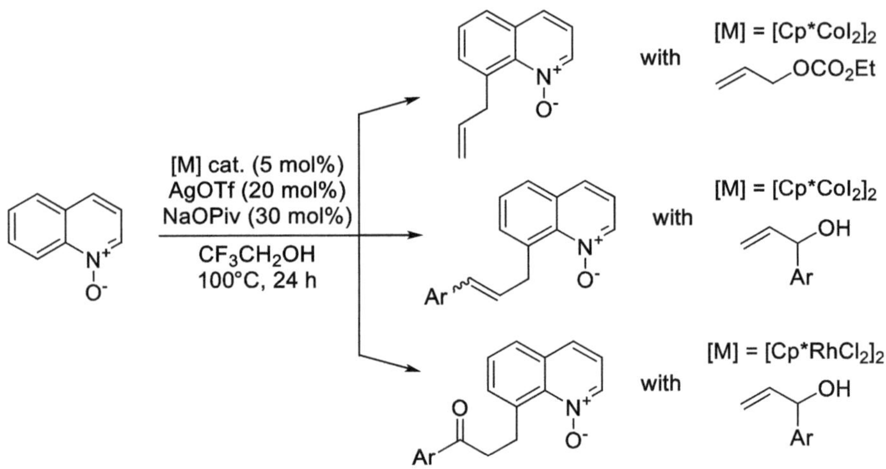

Scheme 42 Rh- and Co-catalyzed C8–H alkylation of quinoline *N*-oxides

8-(3-oxoalkyl)quinoline *N*-oxides were formed. A mechanism was proposed that accounted for the observed chemodivergence with the rhodium and cobalt catalysts. With cobalt, the β-oxygen/hydroxy elimination was operative, whereas with rhodium, β-hydride elimination from the oxygenated carbon of the allylic alcohol was proposed to lead to the formation of 3-oxoalkyl group in C8 position of the quinoline *N*-oxide product.

4 Conclusion

The field of C–H functionalization of heterocyclic *N*-oxides has rapidly expanded in the past decade to include efficient methods of arylation, alkenylation, olefination, alkylation, and halogenation. Although significant progress has been made in the area of C2–H functionalization of *N*-oxides, functionalization of the distal positions in quinoline *N*-oxides and related *N*-heterocycles remains a challenge. In particular, functionalization of positions other than C2 and C8 has remained largely unaddressed. It is hoped that new methods that allow access to these distal positions in *N*-heterocyclic systems will be developed in the near future.

Acknowledgments Financial support by the Welch Foundation (AX-1788) and the NSF (CHE-1455061) is gratefully acknowledged.

References

1. Yamaguchi J, Yamaguchi AD, Itami K (2012) Angew Chem Int Ed 51:8960
2. Chung P-Y, Blan Z-X, Pun H-Y, Chang D, Chan AS-C, Chui C-H, Tang JC-O, Lam K-H (2015) Future Med Chem 7:947
3. Mfuh AM, Larionov OV (2015) Curr Med Chem 22:2819
4. Hughes G, Bryce MR (2005) J Mater Chem 15:94
5. Kimyonok A, Wang XY, Weck M (2006) Polym Rev 46:47
6. Coperet C, Adolfsson H, Khuong T-AW, Yudin AK, Sharpless KB (1998) J Org Chem 63:1740
7. Larionov OV, Stephens D, Chavez G, Mfuh A, Arman H, Naumova A, Skenderi B (2014) Org Biomol Chem 12:3026
8. Amir E, Rozen S (2006) Chem Commun 2006:2262
9. Fife WK (1983) J Org Chem 48:1375
10. Briet N, Brookes MH, Davenport RJ, Galvin FCA, Gilbert PJ, Mack SR, Sabin V (2002) Tetrahedron 58:5761
11. Londregan AT, Jennings S, Wei L (2011) Org Lett 13:1840
12. Medley JW, Movassaghi M (2009) J Org Chem 74:1341
13. Wengryniuk SE, Weickgenannt A, Reiher C, Strotman NA, Chen K, Eastgate MD, Baran PS (2013) Org Lett 15:792
14. Larionov OV, Stephens D, Mfuh A, Chavez G (2014) Org Lett 16:864
15. Stephens DE, Chavez G, Valdes M, Dovalina M, Arman HD, Larionov OV (2014) Org Biomol Chem 12:6190
16. Seregin IV, Gevorgyan V (2007) Chem Soc Rev 36:1173
17. Colby DA, Bergman RG, Ellman JA (2010) Chem Rev 110:624
18. Wencel-Delord J, Droge T, Liu F, Glorius F (2011) Chem Soc Rev 40:4740
19. Zhao D, You J, Hu C (2011) Chem Eur J 17:5466
20. Bull JA, Mousseau JJ, Pelletier G, Charette AB (2012) Chem Rev 112:2642
21. Rossi R, Bellina F, Lessi M, Manzini C (2014) Adv Synth Catal 356:17
22. Yan G, Borah AJ, Yang M (2014) Adv Synth Catal 356:2375
23. Davies HML, Morton D (2016) J Org Chem 81:343
24. Iwai T, Sawamura M (2015) ACS Catal 5:5031
25. Sharma R, Thakur K, Kumar R, Kumar I, Sharma U (2015) Catal Rev 57:345
26. Stephens DE, Larionov OV (2015) Tetrahedron 71:8683
27. Campeau L-C, Bertrand-Laperie M, Leclerc J-P (2008) J Am Chem Soc 130:3276
28. Zhao H, Wang R, Chen P, Gregg BT, Hsia MM, Zhang W (2012) Org Lett 14:1872
29. Demory E, Farran D, Baptiste B, Chavant PY, Blandin V (2012) J Org Chem 77:7901
30. Liu W, Li Y, Xu B, Kuang C (2013) Org Lett 15:2342
31. Fujiwara Y, Moritani I, Danno S, Asano R, Teraniski S (1969) J Am Chem Soc 91:7166
32. Liu W, Li Y, Wang Y, Kuang C (2013) Org Lett 15:4682
33. Liu W, Yu X, Li Y, Kuang C (2014) Chem Commun 50:9291
34. Li G, Jia C, Sun K (2013) Org Lett 15:5198
35. Wu Z, Song H, Cui X, Pi C, Du W, Wu Y (2013) Org Lett 15:1270
36. Zhu C, Yi M, Wei D, Chen X, Wu Y, Cui X (2014) Org Lett 16:1840
37. Li G, Jia C, Sun K, Lu Y, Zhao F, Zhou K, Wu H (2015) Org Biomol Chem 13:3207
38. Yu H, Dannenberg CA, Li Z, Bolm C (2016) Chem Asian J 11:54
39. Campeau L-C, Rousseaux S, Fagnou K (2005) J Am Chem Soc 127:18020
40. Campeau L-C, Schipper DJ, Fagnou K (2008) J Am Chem Soc 130:3266
41. Campeau L-C, Fagnou K (2011) Org Synth 88:22
42. Leclerc J-P, Fagnou K (2006) Angew Chem Int Ed 45:7781; (2006) Angew Chem 118:7945
43. Schipper DJ, El-Salfiti M, Whipp CJ, Fagnou K (2009) Tetrahedron 65:4977
44. Lafrance M, Blaquiere N, Fagnou K (2007) Eur J Org Chem 2007:811

45. Zhao L, Tsukano C, Kwon E, Takemoto Y, Hirama M (2013) Angew Chem Int Ed 52:1722; (2013) Angew Chem 125:1766
46. Do H-Q, Daugulis O (2007) J Am Chem Soc 129:12404
47. Do H-Q, Khan RMK, Daugulis O (2008) J Am Chem Soc 130:15185
48. Zhao D, Wang W, Yang F, Lan J, Yang L, Gao G, You J (2009) Angew Chem Int Ed 48:3296; (2009) Angew Chem 121:3346
49. Do H-Q, Daugulis O (2010) Org Lett 12:2517
50. Makida Y, Ohmiya H, Sawamura M (2012) Angew Chem Int Ed 51:4122
51. Kanyiva KS, Nakao Y, Hiyama T (2007) Angew Chem Int Ed 46:8872; (2007) Angew Chem 119:9028
52. Cho SH, Hwang SJ, Chang S (2008) J Am Chem Soc 130:9254
53. Huestis MP, Fagnou K (2009) Org Lett 11:1357
54. Xi P, Yang F, Qin S, Zhao D, Lan J, Gao G, Hu C, You J (2010) J Am Chem Soc 132:1822
55. Wang Z, Song F, Zhao Y, Huang Y, Yang L, Zhao D, Lan J, You J (2012) Chem Eur J 18:16616
56. Gong X, Song G, Zhang H, Li X (2011) Org Lett 13:1766
57. Yamaguchi AD, Mandal D, Yamaguchi J, Itami K (2011) Chem Lett 40:555
58. Madal D, Yamaguchi AD, Yamaguchi J, Itami K (2011) J Am Chem Soc 133:19660
59. Wang Z, Li K, Zhao D, Lan J, You J (2011) Angew Chem Int Ed 50:5365
60. Wu Z, Pi C, Cui X, Bai J, Wu Y (2013) Adv Synth Catal 355:1971
61. Xiao B, Liu Z-J, Liu L, Fu Y (2013) J Am Chem Soc 135:616
62. Wu X, See JWT, Xu K, Hirano H, Roger J, Hierso J-C, Zhou J (2014) Angew Chem Int Ed 53:13573
63. Jha AK, Jain N (2016) Chem Commun 52:1831
64. Hwang H, Kim J, Jeong J, Chang S (2014) J Am Chem Soc 136:10770
65. Kwak J, Kim M, Chang S (2011) J Am Chem Soc 133:3780
66. Gwon D, Hwang H, Kim HK, Marder SR, Chang S (2015) Chem Eur J 21:17200
67. Jeong J, Patel P, Hwang H, Chang S (2014) Org Lett 16:4598
68. Zhang X, Qi Z, Li X (2014) Angew Chem Int Ed 53:10793
69. Sharma U, Park Y, Chang S (2014) J Org Chem 79:9899
70. Li Y, Liu S, Qi Z, Qi X, Li X, Lan Y (2015) Chem Eur J 21:10131
71. Barsu N, Sen M, Premkumar JR, Sundararaju B (2016) Chem Commun 52:1338
72. Shitbata T, Matsuo Y (2014) Adv Synth Catal 356:1516
73. Stephens DE, Lakey-Beitia J, Atesin AC, Atesin TA, Chavez G, Arman HD, Larionov OV (2015) ACS Catal 5:167
74. Shin K, Park S-W, Chang S (2015) J Am Chem Soc 137:8584
75. Stephens DE, Lakey-Beitia J, Chavez G, Ilie C, Arman HD, Larionov OV (2015) Chem Commun 51:9507
76. Stephens DE, Lakey-Beitia J, Burch JE, Arman HD, Larionov OV (2016) Chem Commun 52:9945
77. Stephens DE, Nguyen VT, Chhetri B, Clark ER, Arman HD, Larionov OV (2016) Org Lett 18:5808
78. Yu S, Wan B, Li X (2015) Org Lett 17:58
79. Chen X, Cui X, Wu Y (2016) Org Lett 18:2411
80. Chen X, Cui X, Wu Y (2016) Org Lett 18:3722
81. Suresh R, Kumaran RS, Senthilkumar V, Muthusubramanian S (2014) RSC Adv 4:31685
82. Kalsi D, Laskar RA, Barsu N, Premkumar JR, Sundararaju B (2016) Org Lett 18:4198

Top Heterocycl Chem (2017) 53: 85 110
DOI: 10.1007/7081_2017_2

Published online: 13 April 2017

Recent Advances in Cycloaddition Reactions of Heterocyclic *N*-Oxides

Rafał Loska

Abstract 1,3-Dipolar cycloaddition of *N*-oxides of azines and azoles has become a reliable and versatile synthetic method of preparation of highly functionalized nitrogen heterocycles. Mechanisms of cycloaddition of *N*-oxides are outlined, including various reaction pathways available for the initial five-membered cycloadducts. Cycloaddition to multiple carbon–carbon, carbon–nitrogen, and carbon–sulfur bonds is discussed, with particular emphasis on modern methods of selective C-2-functionalization of the heteroaromatic ring.

Keywords Alkenes • Alkynes • Amides • Aromatic compounds • Azines • Azoles • Dipolar cycloaddition • Fluoroorganic compounds • Isocyanates • Isoxazolidines • Isoxazolines • Nitrogen heterocycles • *N*-oxides • Reaction mechanisms • Thioketones

Contents

R. Loska (✉)
Institute of Organic Chemistry, Polish Academy of Sciences, Kasprzaka 44/52, 01-224 Warsaw, Poland
e-mail: rafal.loska@icho.edu.pl

1 Introduction

Heterocyclic *N*-oxides are readily available and versatile substrates for preparation of highly functionalized nitrogen heterocycles [1–5]. General and useful synthetic methods exploiting *N*-oxides as starting materials have been developed, such as *cine* [6–12] and other types [13–15] of nucleophilic substitution, metalation with organometallic reagents [16], or, more recently, transition metal-catalyzed C–H activation of *N*-oxides of azines [17–29] and azoles [30–33]. The main focus of this review are useful and general synthetic approaches to the preparation of C-2 functionalized azines and azoles based on 1,3-dipolar cycloaddition that has been developed in recent years. Aromatic rings of *N*-oxides of azines (pyridines, quinolines, etc.) and azoles (imidazoles, thiazoles, etc.) can be viewed as containing in their structure a nitrone moiety, which makes them viable substrates for dipolar cycloaddition to multiple carbon–carbon or carbon–heteroatom bonds (Scheme 1) [34]. Cycloaddition of a dipolarophile to the 1,3-dipole formed by C-2, nitrogen, and oxygen is associated with the formation of two new bonds; thus it may be employed as a method of selective functionalization of the heteroaromatic ring at the position vicinal to azine nitrogen.

2 Heterocyclic *N*-Oxides as 1,3-Dipoles

Involvement of the nitrone structural motif in the aromatic ring of *N*-oxides results in lower activity toward dipolarophiles compared with typical nitrones. The cycloaddition step (Scheme 2, top) may occur as a concerted dipolar cycloaddition, which can be classified as first or second type (normal or neutral electron demand) cycloaddition [35] as *N*-oxides react preferentially with strongly electron-deficient dipolarophiles (or highly active ones, e.g., benzyne). On the other hand, owing to the polar nature of *N*-oxides (nucleophilic, negatively charged oxygen vs positively charged C-2 carbon involved in the aromatic ring), they tend to undergo cycloaddition via a highly polarized transition state or in a stepwise manner via a betaine intermediate, especially with strongly electrophilic partners capable of efficient stabilization of negative charge. Many reactions, which can be formally classified as cycloadditions, in fact occur via initial nucleophilic attack of the *N*-oxide oxygen on the more electron-deficient atom of the multiple bond of dipolarophile, followed by the ring closure of the resultant zwitterion, that is, addition of anion to the ring of the intermediate *N*-alkyl azinium (or azolium) salt. Indeed, such reactions are very

Scheme 1 The structure of azine and azole *N*-oxides

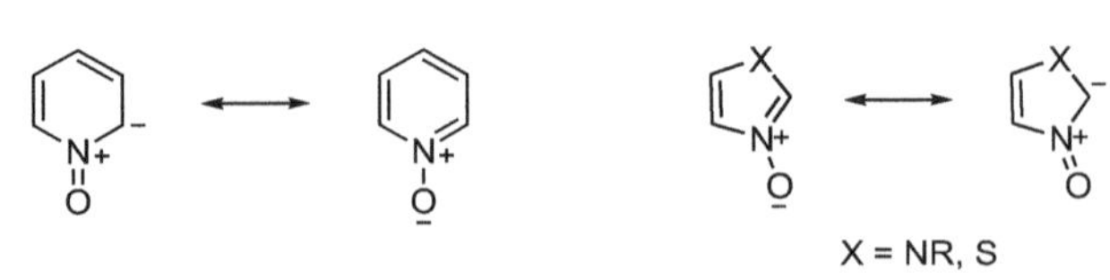

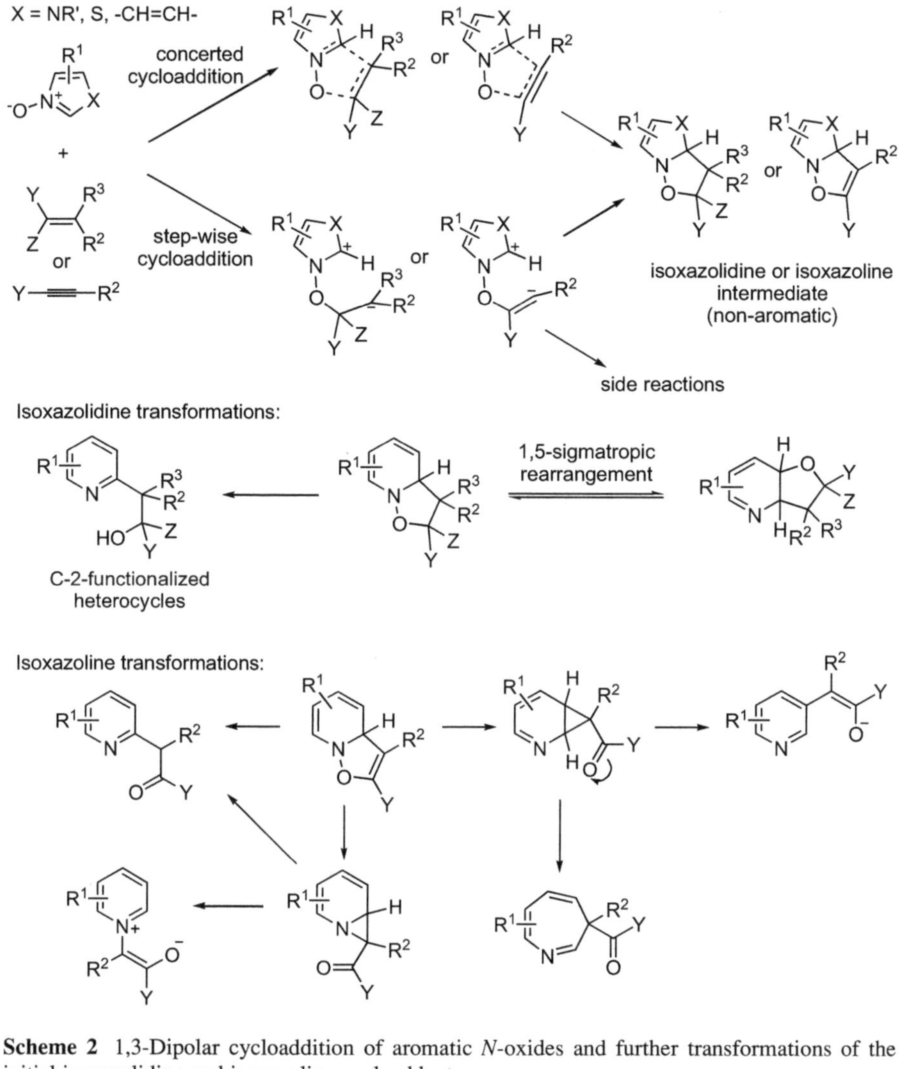

Scheme 2 1,3-Dipolar cycloaddition of aromatic *N*-oxides and further transformations of the initial isoxazolidine and isoxazoline cycloadducts

similar to *cine* substitution in *N*-oxides, in which addition of electrophile activates the oxygen atom and transforms it into a good leaving group, and subsequent attack of nucleophile on the ring and elimination gives substitution product [6]. Formation of a zwitterionic adduct is often followed by other reactions rather than ring closure, leading to the formation of side products such as deoxygenated azines and azoles or pyridones, imidazolones, etc.

Little attention has been given to theoretical calculations concerning cycloaddition involving *N*-oxides as 1,3-dipoles. Simple PM3 calculations performed for a model reaction of pyridine *N*-oxide and isocyanates (R–N=C=O, R=H, Me, Ph) revealed that in the transition state (TS) the formation of the bond between the *N*-oxide oxygen and the central isocyanate carbon is much more advanced than of the bond from isocyanate nitrogen to the C-2 carbon of the pyridine ring [36]. This asynchronous character of the bond-forming process is associated with high partial charges on pyridine nitrogen and isocyanate oxygen and a strong increase of the dipole moment of the TS. Moreover, the calculations predicted the presence of a small saddle just before the TS, corresponding to a betaine intermediate, as well as initial formation of a charge-transfer complex between the two reactants. The existence of such complexes has been earlier confirmed experimentally [37]. Their formation determines the initial orientation of *N*-oxide and dipolarophile and thus regioselectivity of cycloaddition to unsymmetrically substituted pyridine *N*-oxides. By lowering the energy of the reactants with respect to the cycloaddition TS, formation of charge-transfer complexes inhibits reactions of some electron-rich oxides with strongly electron-deficient partners [37, 38]. In contrast to isocyanates, earlier calculations predicted the reaction of pyridine *N*-oxide with electron-deficient allenes to be a fully concerted cycloaddition [39].

Unlike nitrones, the formation of the five-membered cycloadduct in the *N*-oxide cycloaddition is associated with the loss of the resonance stabilization energy of the starting azine or azole ring. Usually, this cycloadduct is not the final, isolable reaction product, but it undergoes further transformations, the driving force of which is restoration of aromaticity of the original heterocycle. Several early examples of reactions proceeding according to different pathways illustrated in Scheme 2 can be found in the review of Ryzhakov and Rodina [34].

The most straightforward path is a proton abstraction from the sp^3 carbon vicinal to nitrogen, accompanied by a heterolytic cleavage of the N–O bond. This process is typical for cycloadducts formed from azole *N*-oxides and is the basis for synthetically useful reactions that introduce various groups at C-2 position of azine/azole ring. For azines, this pathway is not necessarily the dominating one as the isoxazolidine intermediate can undergo a 1,5-sigmatropic rearrangement with the cleavage of the N–O bond. The resulting furopyridines are often isolable products. For isoxazoline cycloadducts, resulting from cycloaddition to triple bonds, other pathways have been observed, such as the N–O cleavage associated with rearrangement to a 1,2-dihydropyridine-fused aziridine intermediate. Depending on the nature of the substituents at the initial triple bond, the aziridine ring opens to form either a 2-substituted azine or an azinium ylide. Alternatively, rearrangement to a 2,3-dihydropyridine-fused cyclopropane may occur, followed by the heterolytic ring opening to a 3-substituted pyridine or electrocyclic opening to azepine.

The multitude of transformations available for cycloadducts of *N*-oxides and dipolarophiles (Scheme 2 shows only the most important ones) perhaps explains why for many years *N*-oxide cycloaddition had only moderate application in synthetic organic chemistry. The key feature of the processes developed recently and described in detail in the following sections is an exclusive or strongly predominating occurrence of only one possible pathway for a fairly wide range of substrates.

3 Cycloaddition to Carbon–Carbon Multiple Bonds

Dimethyl acetylenedicarboxylate (DMAD) was among the earliest dipolarophiles employed in cycloaddition with aromatic *N*-oxides. For example, Hamana and coworkers investigated its cycloaddition with a quinoline *N*-oxide derivative which serves well to illustrate the complexity of azine *N*-oxides cycloadditions [40]. Reaction of 3-bromo-4-methoxyquinoline *N*-oxide with DMAD (dioxane, rt, or reflux) leads to three products which presumably originate from the initial isoxazoline cycloadduct (Scheme 3). Only one of these products can be rationalized by a simple N–O bond cleavage. For the other two (*N*-azinium ylide and *N*-vinylquinoline derivative), isomerization to an aziridine intermediate has to be invoked. Their formation is possible via subsequent C–C bond cleavage, whereas C–N bond fissure can be an alternative pathway of formation of a 2-substituted quinoline. A similar reaction course has been observed for the cycloaddition of DMAD to 1,2-dimethylbenzimidazole 3-oxide [41].

Abramovitch investigated the reactions between azine *N*-oxides and an electron-poor alkyne, phenyl cyanoacetylene (PhC≡CCN), which gives mixtures of the expected 2-alkylation products together with α-cyanophenacylpyridinium ylides [42, 43]. While the formation of *N*-ylides is analogous to those observed by Hamana in the reaction of DMAD, the formation of a 3-pyridyl ylide can be rationalized by a rearrangement to a cyclopropane-fused 2,3-dihydropyridine, followed by cyclopropane ring opening (Scheme 4). According to expectation, this process is avoided in the case of 3,5-lutidine *N*-oxide, which gives a 2-alkylation product in high yield.

In contrast to the unselective processes described above, the reaction of pyridine *N*-oxide with dibenzoylacetylene proceeds readily at rt to give the C-2 substitution product in high yield [44].

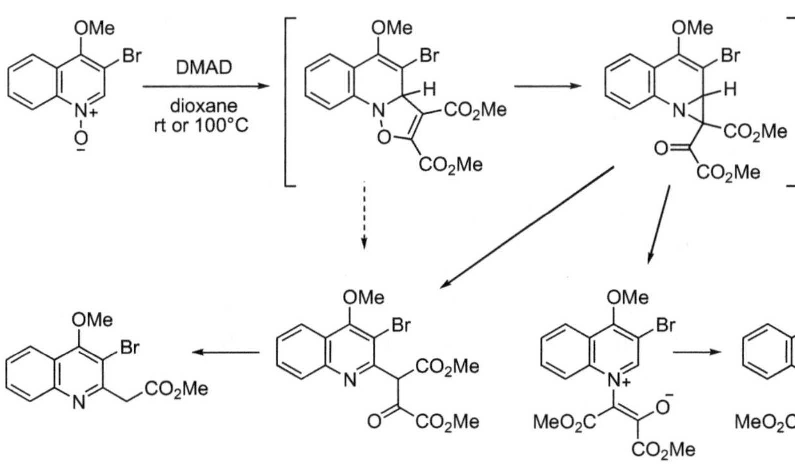

Scheme 3 Cycloaddition of DMAD and quinoline *N*-oxide derivative

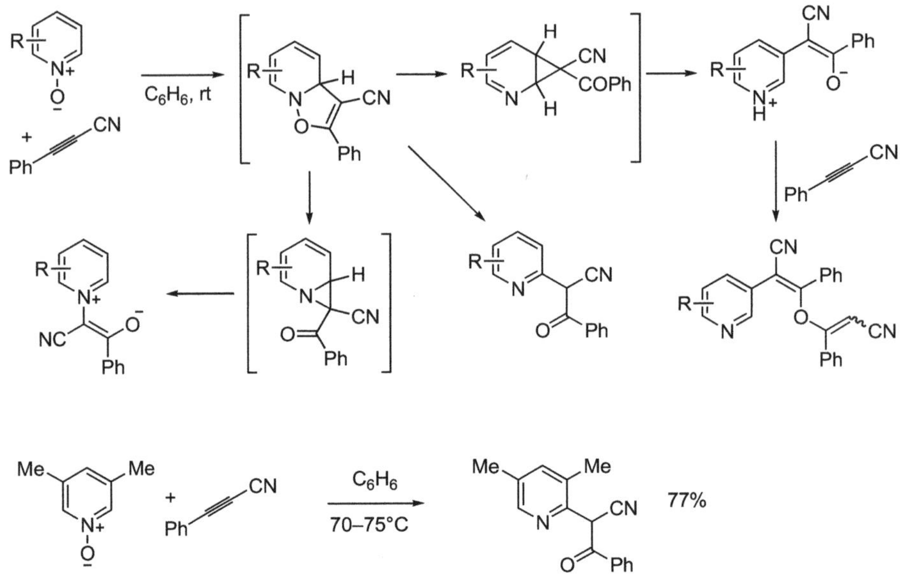

Scheme 4 Cycloaddition of an electron-deficient alkyne with pyridine *N*-oxide derivatives

Selective cycloaddition of a highly functionalized purine *N*-oxide with 3-phenyl-2-propynenitrile has been utilized to introduce selectively a 1-cyano-2-hydroxy-2-phenylvinyl substituent into the heterocyclic ring in the synthesis of modified purine nucleosides [45].

Benzyne is another highly active dipolarophile containing a triple carbon–carbon bond and reacting readily with *N*-oxides. A selective method for 3-hydroxyarylation of pyridines based upon cycloaddition of *N*-oxides with benzyne, generated from *ortho*-trimethylsilylaryl triflates in the presence of CsF, has been described by Larock [46]. Regioselectivity of this reaction has been explained by a sigmatropic rearrangement as the dominating pathway of the decomposition of the initial cycloadduct, even though it occurs through a cyclopropane-fused 2,3-dihydropyridine intermediate, in which both rings are nonaromatic (Scheme 5).

More recently, this reaction has been revisited by Liu and coworkers, who found that regioselectivity 2- vs 3-arylpyridine can be tuned by the variation of the medium polarity and excess of basic species (*N*-oxide and fluoride source) with respect to the benzyne precursor [47]. According to the updated mechanism of cycloaddition to benzyne, the initial cycloadduct and the cyclopropane rearranged product are in equilibrium (Scheme 6). Basic species present in the reaction mixture extract the proton from the C-2 position with concomitant cleavage of the N–O bond and rearomatization of the pyridine moiety. At low concentration of bases, ring opening of the cyclopropane intermediate is favored. It is facilitated by the presence of lithium cations or electron-poor alkynes (such as ethyl propiolate), which add to the ketone oxygen to form ethers of 3-hydroxyarylpyridines.

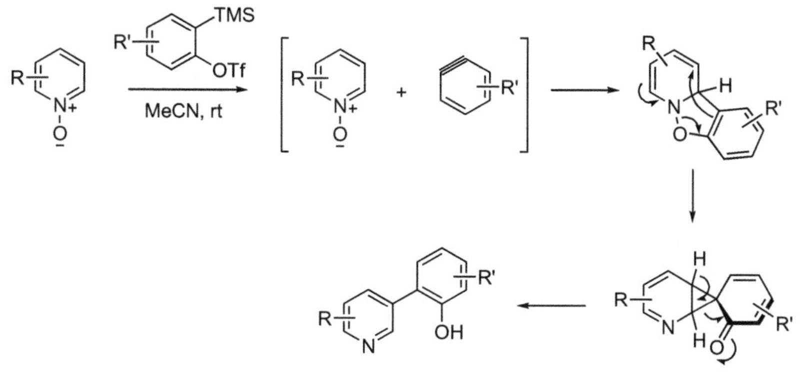

Scheme 5 Preparation of 3-hydroxyarylpyridines via cycloaddition of pyridine *N*-oxides with benzyne

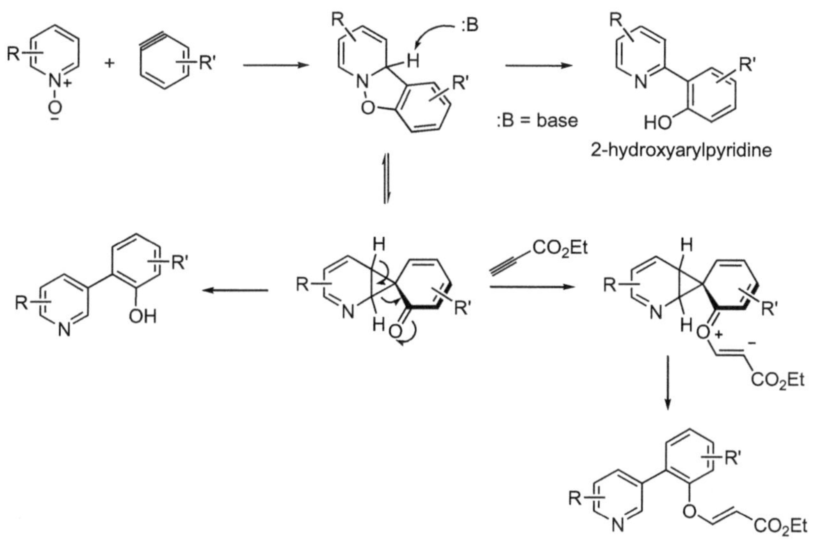

Scheme 6 An updated mechanism for the reaction between pyridine *N*-oxides and benzyne [47]

Cycloaddition of imidazole *N*-oxides with DMAD gives the expected products of the isoxazoline ring opening [48]. An analogous behavior has been reported for 1-methylbenzimidazole 3-oxide [41, 49] and dihydroimidazole *N*-oxide [50]. More recently, Mloston and coworkers reported that *N*-oxides of imidazoles bearing an electron-withdrawing group at C-4 (ester, acetyl, carboxamide) react with DMAD to give the expected 2-oxobutanedioates, but the products readily undergo addition of water and oxalyl cleavage to (imidazol-2-yl)acetates [51, 52].

Reactions of azine *N*-oxides with acrylates [53, 54] and acrylonitriles [55] have been attempted already since the 1960s. These reactions led to 3-(2-hetroaryl)-2-hydroxy derivatives of propionic acids, formed via dipolar cycloaddition and the cycloadduct ring opening (Scheme 7). The structure of the products indicates regioselectivity of cycloaddition opposite to the one expected for initial nucleophilic addition of the oxygen atom to the double bond followed by cyclization. After 2000, this reaction has seen renewed interest as a method of selective C-2 alkylation of azines.

A one-pot reaction has been described in which *N*-oxides of isoquinolines are generated in situ by an AgOTf-catalyzed cyclization of 2-alkynylbenzaldoximes and then subjected to the reaction (without isolation) with acrylates, methacrylates, or acrylonitriles at 60°C to give as products hydroxyalkyl-substituted isoquinolines [56]. The structure of the products can be explained by dipolar cycloaddition of *N*-oxide to alkene and subsequent ring opening of the isoxazolidine cycloadduct, although different reaction pathways may be in operation, especially considering the presence of Ag salts in the reaction mixture. A similar cycloaddition mechanism is possible for the reaction described earlier by Wu and coworkers, who generated 4-haloisoquinoline *N*-oxides from 2-alkynylbenzaldoximes upon the action of Br_2 or ICl in the presence of α,β-unsaturated carbonyl compounds, although the authors themselves proposed a different, very complicated mechanism which does not involve an isoxazolidine intermediate [57].

Very recently, two research groups reported on cycloaddition between quinoline *N*-oxides and acrylates as a selective, general, and transition metal-free method of C-2 quinoline functionalization. The group of Sharma disclosed that heating quinoline *N*-oxides with excess of acrylates to 100°C without solvent results in the formation of 2-hydroxy-3-(2-quinolinyl)propionates, with small amounts of 3-(2-quinolinyl)acrylates as side products [58]. The reaction is general for quinoline *N*-oxides with both electron-donating and electron-withdrawing substituents (except NO_2), and it tolerates sensitive functional groups such as acetals. Isotopic labeling experiments showed that the oxygen atom in the hydroxy group of the products comes from the *N*-oxide substrate. *N*-Oxides of electron-deficient quinolines were less reactive than electron-rich ones. This result indicates that the key reaction step is a normal electron demand dipolar cycloaddition.

Wang and coworkers described the synthesis of 2-alkenylquinolines in the reaction of quinoline *N*-oxides with acrylates and styrenes [59]. The reaction requires quite harsh conditions (DMSO, AcOH, 120°C) and works only for monosubstituted olefins, but it is efficient for a variety of quinoline derivatives and styrenes of different electronic character. Isolation of ethyl 2-hydroxy-3-

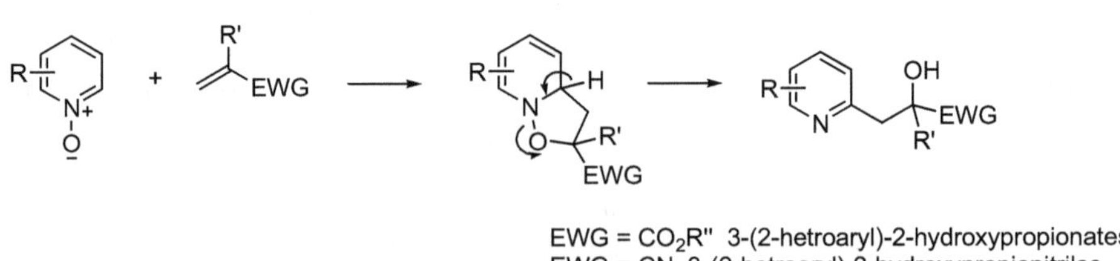

Scheme 7 Cycloaddition of azine *N*-oxides with acrylates and acrylonitriles

(2-quinolinyl)propionate from a reaction performed for a shorter period of time indicates that it is an example of dipolar cycloaddition (Scheme 8). The role of AcOH additive is facilitation of the N–O bond cleavage and dehydration.

Cycloaddition of acrylates with imidazole *N*-oxides has been reported to display opposite regioselectivity than that of *N*-oxides of azines. Imidazole *N*-oxides upon heating above 100°C with ethyl acrylate gave ethyl 3-hydroxy-2-(2-imidazolyl) propanoates [60]. These products were formed in moderate yields, and only for 4,5-diphenylimidazole *N*-oxides; other imidazole *N*-oxides gave only decomposition products. 3-Hydroxy-2-(2-imidazolyl)propanoates could be dehydrated in the presence of $Pd(OAc)_2$ to 2-vinylimidazole derivatives.

The same group investigated reactions of hydrates of 4-phenylimidazole-3-oxides with a strongly electrophilic alkene, 2,2-bis(trifluoromethyl)ethene-1,1-dicarbonitrile in CH_2Cl_2, and methanol [61]. The reaction occurred within minutes at room temperature to provide mixtures of imidazol-2-ones and imidazol-2-ylidenemalononitriles, together with hexafluoroacetone hydrate. These results prompted the authors to propose a stepwise cycloaddition mechanism. A zwitterionic intermediate (Scheme 9), which is the product of nucleophilic attack of the *N*-oxide oxygen on the double bond of alkene, may undergo an isoxazolidine ring closure to a formal cycloadduct, or it may suffer attack of a water molecule on the activated C-2 position of the ring, leading to the formation of imidazol-2-one. This reaction pathway can be classified as a *cine* nucleophilic substitution of hydrogen and it is dominant in CH_2Cl_2, in which water present in the substrate is closely bound to highly polar *N*-oxide. In the former pathway, which predominates in MeOH, cleavage of the N–O isoxazolidine bond and rearomatization, followed by a retro-aldol loss of difluoroacetone, provides a malononitrile derivative, in its more stable tautomer with an exocyclic double bond. The same malononitrile derivatives could be obtained exclusively if carefully dried chloroform solutions of *N*-oxides were employed in the reaction.

The outcome of cycloaddition between azine *N*-oxides and an electron-poor allene (dimethyl 2,3-pentadienedioate) has been reported to depend strongly on the structure of azine [62]. Some pyridine derivatives (2-Ph, 3,5-Cl_2, 3,5-Br_2, 3-CO_2Me) react smoothly to give stable products of 1,5-sigmatropic rearrangement

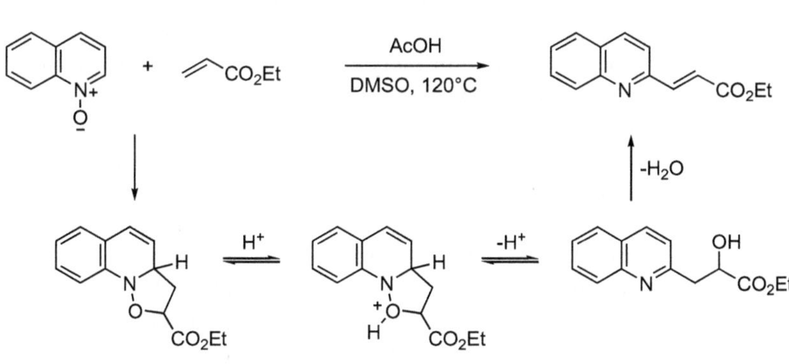

Scheme 8 Alkenylation of quinolines

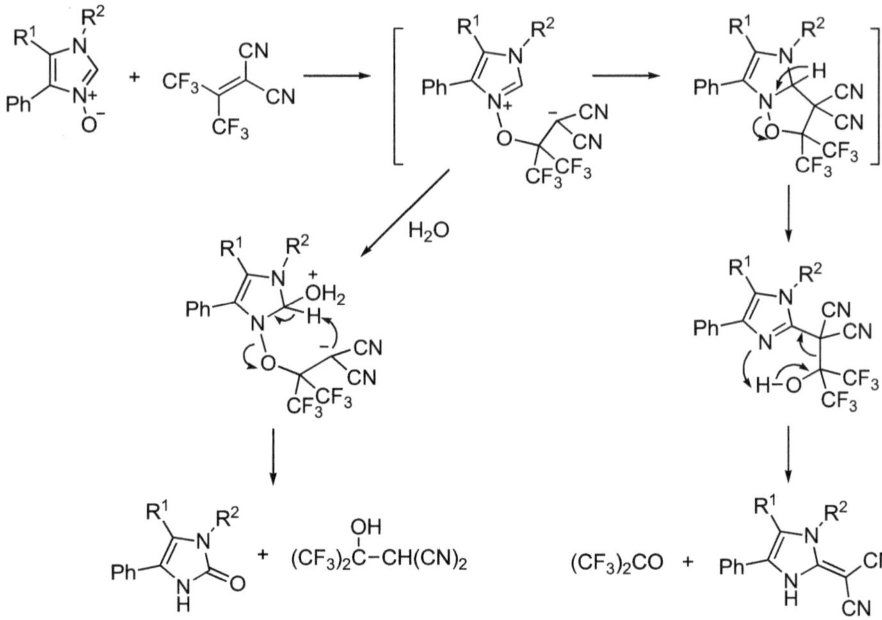

Scheme 9 Reaction of imidazoles *N*-oxides with 2,2-bis(trifluoromethyl)ethene-1,1-dicarbonitrile

(Scheme 10). The reaction is stereoselective in terms of the exocyclic double bond configuration, which implies a concerted cycloaddition mechanism. Simple theoretical calculations (MNDO) allowed to explain its regioselectivity in terms of the FMO theory and to classify it as a normal-type cycloaddition with the dominant interaction between the HOMO of dipole and the LUMO of dipolarophile. On the other hand, cycloaddition of quinoline *N*-oxide is followed by ring opening of the initial cycloadduct to give a 2-vinylated product, which is easily understood as a 1,5-rearrangement would produce a quinodimethane-like structure. *N*-Oxides of pyridine and picolines gave complex mixtures of products.

Cycloaddition with azine *N*-oxides of relatively low resonance stabilization energy may lead to stable, isolable primary cycloadducts, as described, for example, for addition of quinoline *N*-oxide to 1,4-epoxy-1,4-dihydronaphthalene, quinoxaline 1,4-dioxide to maleimide [34], or quinazoline 3-oxides to *N*-methylmaleimide and vinyl phenyl sulfone [63]. In some cases, a cycloadduct formed from quinazoline 3-oxide may undergo unusual reactions, as reported recently for the reaction with alkylidenecyclopropanes in the presence of water and Cu salts (Scheme 11) [64]. The isoxazolidine adduct formed in the dipolar cycloaddition step does not undergo a typical scission of the N–O bond, but rather a Cu-promoted addition of water to the six-membered heterocyclic ring resulting in its cleavage. Final proton shift from the benzylic position to the nitrogen atom provides spirocyclic isoxazolines. The feasibility of this process may be explained

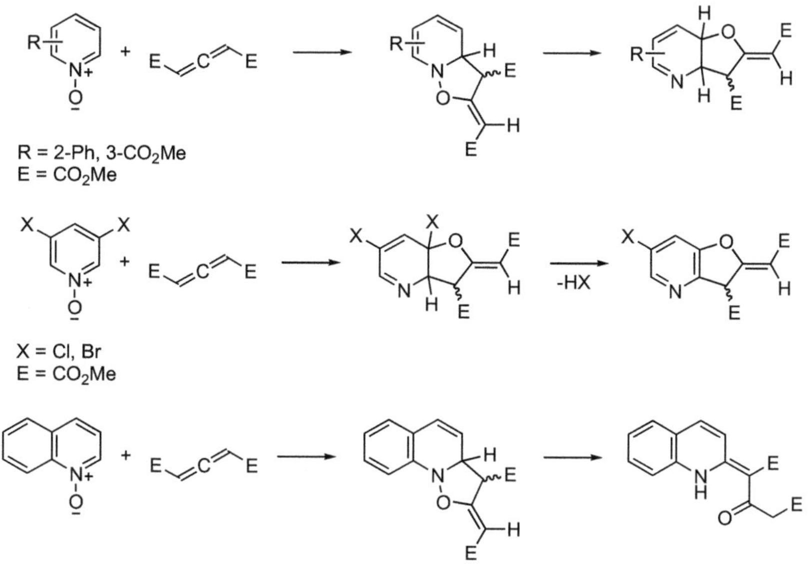

Scheme 10 Reactions of *N*-oxides of azines with electron-poor allenes

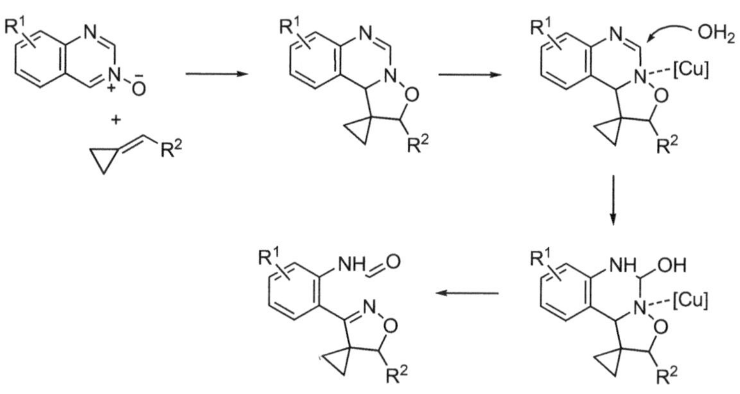

Scheme 11 Cu-catalyzed cycloaddition of quinazoline 3-oxides and alkylidenecyclopropanes followed by heterocyclic ring opening [64]

by relatively low resonance energy of the quinazoline heterocyclic ring which results in relatively high stability of the isoxazolidine intermediate, which survives long enough to be coordinated by copper catalyst. Alternatively, it is possible that the Cu cation is already coordinated to *N*-oxide prior to the cycloaddition step.

Dipolar cycloaddition reactions of fluoroalkenes [65–67] provide new methods of preparation of heterocycles with fluorinated groups, which are substances of increasing practical importance [68–73]. Cycloaddition of pyridine *N*-oxides with hexafluoropropene (C_3F_6, HFP) has been reported already in the 1960s by Mailey and Ocone as a reaction proceeding in an autoclave at elevated temperature and pressure to give 2-(1,2,2,2-tetrafluoroethyl)pyridines. The structure of the products, as well as detection of CO_2 and difluorophosgene among the gaseous products of the reaction, prompted the authors to propose a mechanism in which the 4,5,5-trifluoro-4-trifluoromethylisoxazolidine adducts undergo heterolytic N–O bond cleavage to form α,α-difluoroalcohols, which in turn undergo a retro-aldol loss of $CF_2{=}O$ (Scheme 12) [74, 75].

Much later, it was demonstrated that cycloaddition of *N*-oxides and HFP could be performed using standard laboratory glassware (HFP condensed in a glass pressure tube) at room temperature (Scheme 13) [76]. Moreover, it is a general method of selective fluoroalkylation of nitrogen heterocycles as *N*-oxides of azines, imidazoles, and thiazoles are all viable substrates, as well as olefins such as HFP, 2*H*-pentafluoropropene (PFP), or chlorotrifluoroethylene ($CF_2{=}CFCl$). In one case (quinoline), fluorinated isoxazolidine intermediate could be characterized.

Cycloaddition to perfluoroalkenes proceeds with complete regioselectivity, that is, the *N*-oxide oxygen atom attacks exclusively the CF_2 group of fluoroalkene. Pyridine *N*-oxides with electron-withdrawing groups at C-3 give predominantly 2,3-substituted products, unlike 3-picoline *N*-oxide which gives mainly the less sterically hindered 2,5-substituted pyridine. Such regioselectivity may be associated with a stepwise mechanism of cycloaddition of highly electron-deficient dipolarophiles such as nitrilium cations [77]. Considering that terminal difluoroalkenes are very prone to undergo nucleophilic addition [78], one might suspect a stepwise mechanism for cycloaddition of perfluoroalkenes with *N*-oxides. On the other hand, a stepwise mechanism seems unlikely as the intermediate carbanion resulting from nucleophilic addition would undergo very facile β-elimination of fluoride. This process would lead to various side products which have never been observed.

Investigation of cycloaddition between HFP and quinoline *N*-oxide under various conditions (Scheme 14) led to the observation that tetrafluoroethyl-substituted azines are not the only possible products and to the formulation of a revised mechanism for this reaction [79]. The reaction performed in DMF/D_2O mixture

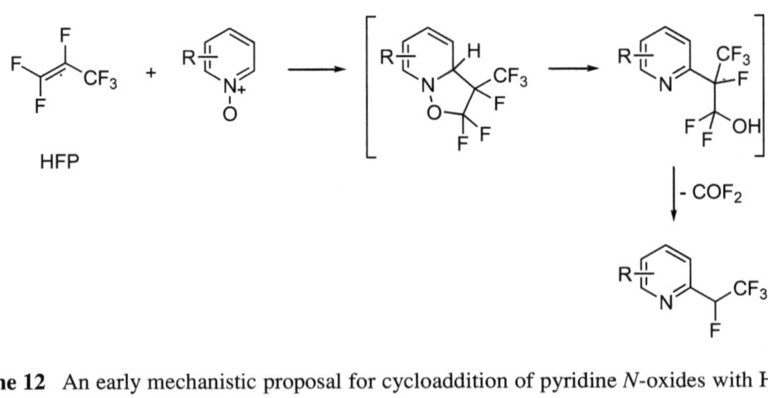

Scheme 12 An early mechanistic proposal for cycloaddition of pyridine *N*-oxides with HFP

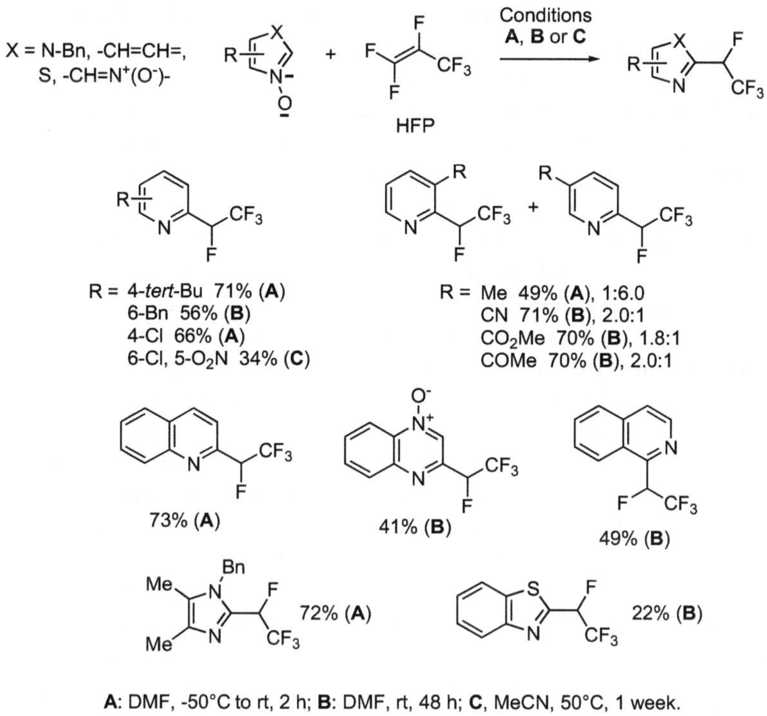

Scheme 13 Cycloaddition of HFP with azine and azole *N*-oxides producing tetrafluoroethyl derivatives of nitrogen heterocycles

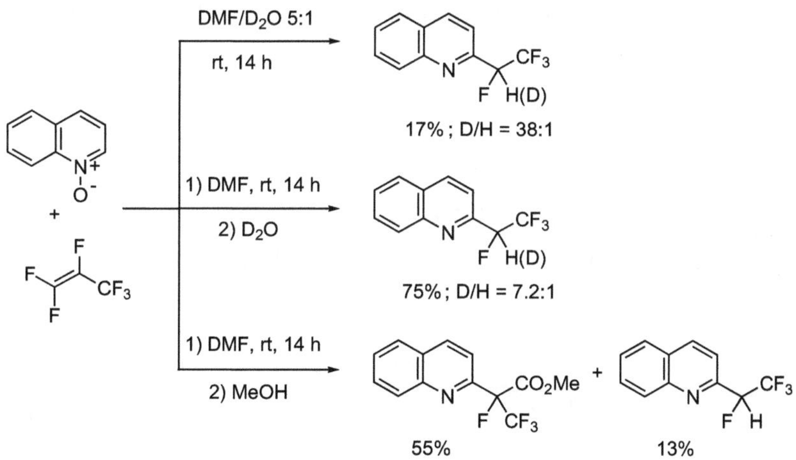

Scheme 14 Influence of D_2O and MeOH on reactions of quinoline *N*-oxide with HFP

instead of DMF alone leads to quinoline with a deuterated fluoroalkyl group. Addition of D_2O after 14 h to the reaction conducted in dry DMF also resulted in mostly deuterated product, D/H 7.2:1. These observations could be explained assuming that decomposition of the intermediate isoxazolidine or α,α-difluoroalcohol occurs only in the presence of water. However, further experiments showed that cycloaddition of quinoline *N*-oxide and HFP performed either in the presence of methanol or with addition of MeOH at the end gave mainly or exclusively methyl ester of 2-(2-quinolinyl)perfluoropropionic acid. This indicates that the reaction proceeds via an acyl fluoride intermediate, which accumulates in the reaction mixture when no protic nucleophile (MeOH, H_2O, etc.) is present (Scheme 15). The existence of acyl fluoride as an intermediate in cycloaddition reactions with fluoroalkenes has been later confirmed by NMR spectroscopy (vide infra). Its hydrolysis gives carboxylic acids which undergo facile decarboxylation, and reactions with other nucleophiles provide derivatives of 2-heteroarylcarboxylic acids. Hydrolysis of methyl esters with, for example, $LiOH \cdot H_2O$ leads directly to tetrafluoroethyl derivatives.

Based on the revised mechanism, a three-component reaction has been developed in which highly functionalized azines and azoles with partially fluorinated groups are assembled from *N*-oxides, perfluoroalkenes and alcohols, amines, or thiols (Scheme 16) [79]. Some amines are nucleophilic enough to react directly with perfluoroalkenes according to nucleophilic addition–elimination mechanism. Fortunately, they can be added to the reaction mixture after the acyl fluoride has been formed in the first cycloaddition step between *N*-oxide and alkene (Scheme 17).

In cycloaddition of *N*-oxides with perfluoroalkenes, the two terminal fluorine atoms are crucial for regioselectivity and exclusive formation of acyl fluoride. This fact inspired the development of an even more general synthetic protocol in which various alkenes containing only two or even one fluorine atom react with *N*-oxides and nucleophiles to produce 2-functionalized azines and azoles with primary,

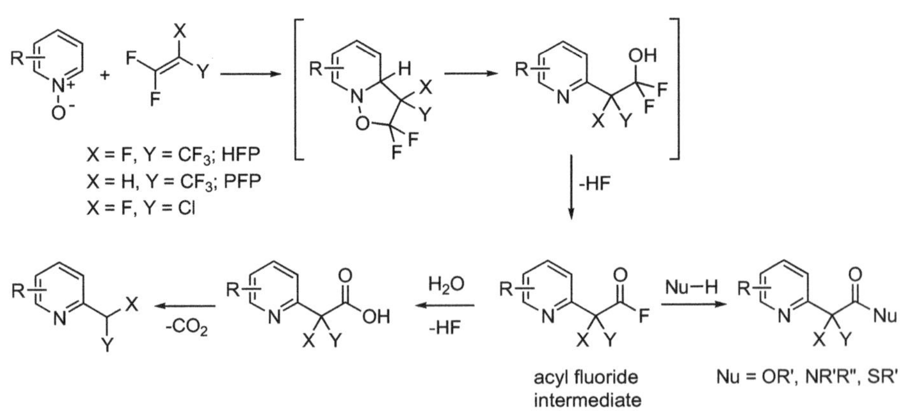

Scheme 15 The revised mechanism of cycloaddition of heterocyclic *N*-oxides with perfluoroalkenes

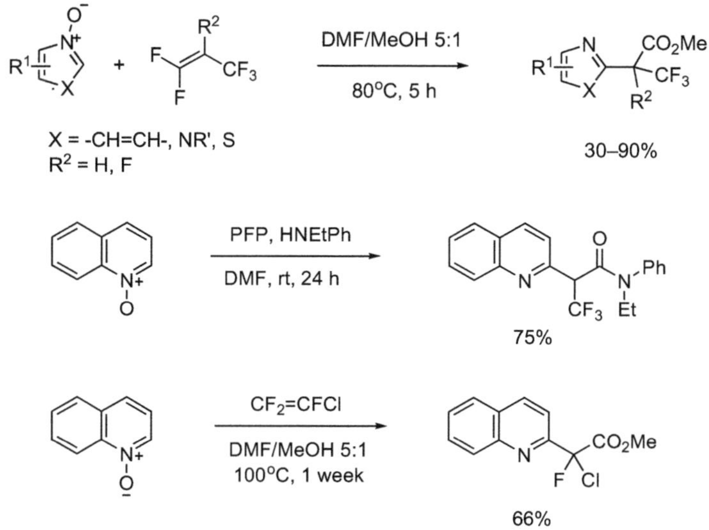

Scheme 16 Examples of reactions of *N*-oxides with perfluoroalkenes in the presence of nucleophiles

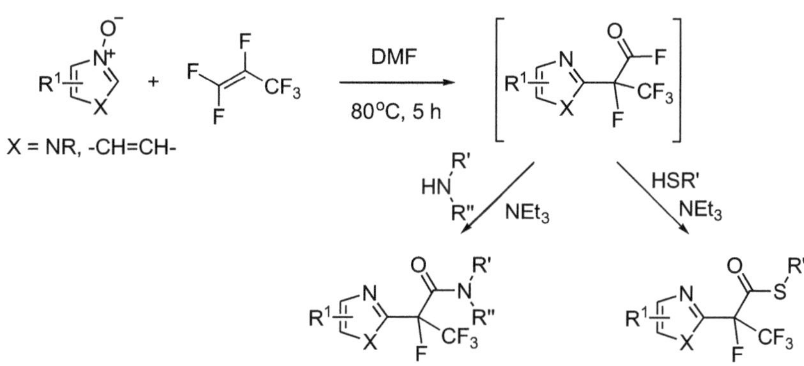

Scheme 17 Two-step reactions of *N*-oxides with HFP and strong nucleophiles

secondary, or tertiary alkyl groups of any structure. Such process would involve the loss of fluorine present in the substrate, but 1,1-difluoroalkenes are nowadays available via a variety of methods, such as Wittig [80–83] and Julia-type [84–86] difluoroolefination of carbonyl compounds, Pd-catalyzed reactions of difluoro-vinylmetal reagents [87–92], and many other reactions [93].

Unfortunately, difluoroalkenes with only fluorine and alkyl substituents at the double bond are unreactive toward aromatic *N*-oxides even under forcing conditions, but 1,1-difluorostyrenes react readily with *N*-oxides of imidazoles, thiazoles, or quinoline and alcohols to give aryl(heteroaryl)acetates (Scheme 18) [94]. The

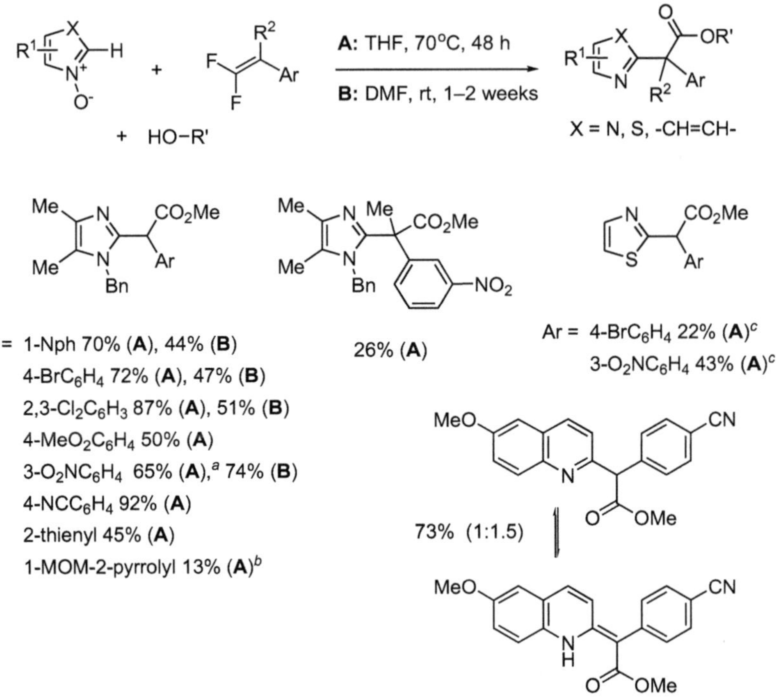

Scheme 18 Synthesis of aryl(heteroaryl)acetates from *N*-oxides, difluorostyrenes, and alcohols. [a] 40°C. [b]93% based on recovered starting material. [c]50°C, 4 days

quinoline-derived product shows preference for the exomethylene tautomeric structure.

Amides of 2-heteroarylacetic acids can be assembled in a one-pot, three-component reaction of *N*-oxides, difluorostyrenes, and various amines (Scheme 19) [95]. Difluorostyrenes are much less electrophilic than HFP or PFP; thus there is usually no need to separate the two steps of the reaction, that is, acyl fluoride-forming 1,3-cycloaddition and reaction with nucleophile. *o*-Phenylenediamine reacts selectively at only one amine group. *N*,*O*-Dimethylhydroxylamine and ammonia (as NH_4Cl) give Weinreb amides and primary amides, respectively. Much better yields of the latter compounds are obtained using hexamethyldisilazane as nucleophile and the source of the amide nitrogen.

Among various types of *N*-oxide substrates, pyridines prove the least reactive, and quinolines and isoquinolines nearly as reactive as imidazoles. Tetrasubstituted 1,1-difluorostyrenes provide imidazoles with all-carbon quaternary centers (including CF_3-substituted ones) connected to the heterocyclic ring, although in lower yields than trisubstituted substrates. According to expectation, a monofluorostyrene gives an aryl(heteroaryl)acetaldehyde in the reaction with quinoline *N*-oxide derivative, but it reacts much more sluggishly than difluorostyrenes [95].

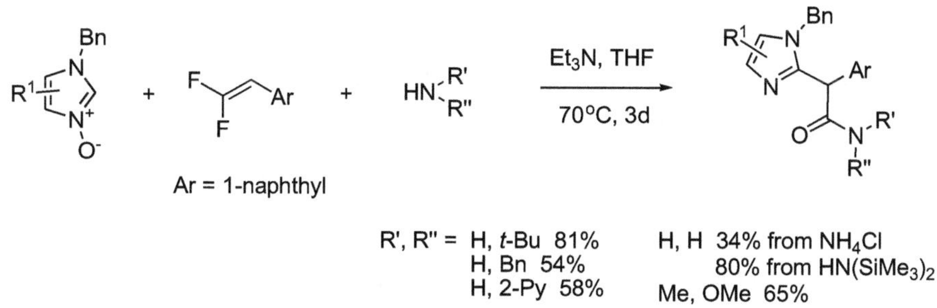

Scheme 19 Three-component synthesis of α-imidazolylacetamides

The mechanism of cycloaddition of *N*-oxides with difluorostyrenes is probably similar to the one discussed for perfluoroalkenes. Reactions of difluorostyrenes are particularly efficient in solvents of moderate polarity such as THF, dioxane, and AcOEt, while reactions in DMF are faster but give lower yields due to formation of multiple unidentified side products. This might be explained by assuming that a stepwise zwitterionic cycloaddition becomes important when polarity of the medium is increased, but concerted cycloaddition (or via a very short-lived zwitterionic intermediate) occurs in THF. This conclusion is further supported by the fact that no imidazol-2-ones were observed as side products, although most of the starting *N*-oxides were used as hydrates, similarly to the reaction in Scheme 9 [61]. Pyridine *N*-oxides with electron-withdrawing groups give less hindered 2,5-substituted products, unlike cycloaddition of nitrilium ions or even HFP [76, 77]. A concerted cycloaddition as the rate-determining step is also in agreement with the much lower activity of tetrasubstituted alkenes.

The further transformations of the initial difluoroisoxazolidine cycloadduct at least in some cases involve an acyl fluoride intermediate. This compound has been observed and characterized by NMR spectroscopy as the sole product when an imidazole *N*-oxide derivative and 2-arylpentafluoropropene were mixed in deuterated THF at 70°C [95]. When *p*-toluidine was added to the reaction mixture prepared in a similar manner, the expected amide was obtained in good yield (Scheme 20). On the other hand, acyl fluoride could not be observed when an alkene with a methyl group instead of CF_3 was used. Accordingly, the reaction of this alkene gave the amide product only if the reaction mixture contained the appropriate amine from the beginning. These observations suggest that the stability of the acyl fluoride intermediate and its ability to accumulate in the reaction mixture containing *N*-oxide and fluoroalkene depends on the electronic nature of the substituents at the α-carbon. When a CF_3 group is not bound to this center, and no nucleophile is present, then acyl fluoride or difluoroalcohol intermediate decomposes quickly, perhaps via a loss of difluorophosgene as suggested already by Mailey and Ocone (see Scheme 12).

Cycloaddition of *N*-oxides and difluorostyrenes allows selective introduction of benzyl-type substituents at C-2 position. Therefore, this reaction can be exploited in the construction of bis(heteroaryl) ligands, if two heteroaryl units with

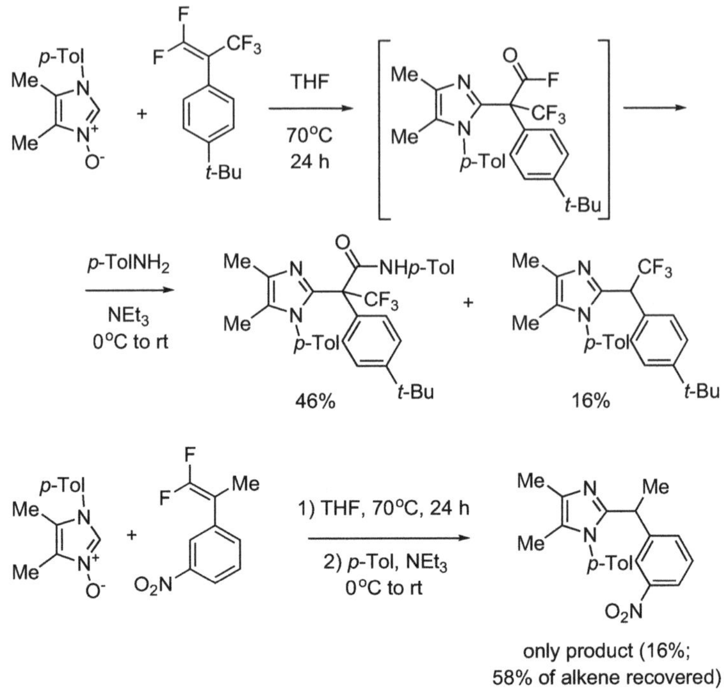

Scheme 20 Quantitative formation of the acyl fluoride intermediate in the reaction of imidazole *N*-oxide with a CF_3-substituted difluorostyrene

appropriately placed azine nitrogens become connected by a central *meso* bridging atom. This concept requires employment of 2-heteroaryl-1,1-difluoroalkenes as cycloaddition partners. In practice, it could be realized with the use of highly electrophilic 2-(2′-heteroaryl)difluoroacrylates [96]. These alkenes cannot be isolated due to their facile hydrolysis, but instead are generated in situ in the presence of *N*-oxides by elimination of HF from 2-heteroaryl-3,3,3-trifluoropropionates with an amine base. The latter compounds are readily available from *N*-oxides of azines and methyl 2*H*-perfluoroisobutyrate via a similar elimination–cycloaddition protocol or cycloaddition with PFP followed by esterification of the acyl fluoride intermediate with MeOH (Scheme 21). Cycloaddition of imidazole *N*-oxides with 2-(2′-heteroaryl)difluoroacrylates produced deoxygenated imidazoles as side products, which suggests a stepwise mechanism and side reactions of the initial zwitterionic adduct.

This synthetic approach allows the assembly of various unsymmetrical ligands of a desired structure from two azine/azole *N*-oxides and a fluorinated synthon which is the source of the *meso* carbon. Bis(heteroaryl)methanes containing two azine units, azine/imidazole or azine/thiazole combinations, were obtained in moderate to good yields (Scheme 22) [96]. Importantly, the method does not rely upon the use of transition metal catalysts which could bind strongly to the final bis

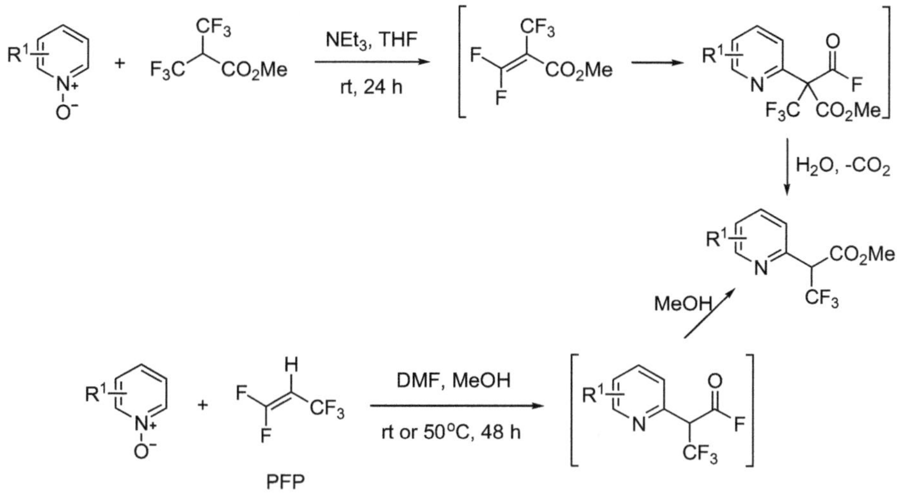

Scheme 21 Synthesis of 2-(2′-heteroaryl)difluoroacrylates

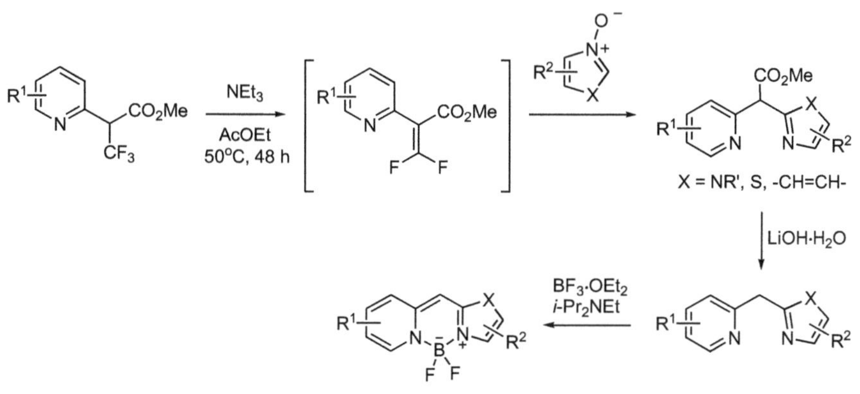

Scheme 22 Synthesis of unsymmetrical bis(heteroaryl)methane derivatives and their BF_2 fluorescent complexes from *N*-oxides via a dipolar cycloaddition approach

(heteroaryl)methanes and therefore be difficult to remove. Some of the azine/imidazole ligands were subjected to hydrolysis, decarboxylation, and subsequent complexation with boron trifluoride to yield labile, but strongly fluorescent unsymmetrical analogues of BODIPY dyes.

4 Cycloaddition to Carbon–Nitrogen Multiple Bonds

The reaction of *N*-oxides of azines with phenyl isocyanate has been investigated since the early 1960s ([97–99], and references therein). In its course, cycloaddition of *N*-oxide to the C=N double bond leads to the formation of cycloadducts, which quickly undergo a 1,5-sigmatropic rearrangement to 2,3-dihydroazine intermediates, which in some cases can be isolated [100–102]. Thermally induced extrusion of CO_2 provides 2-aminoazines. A confirmation of such reaction pathway is the fact that when the starting pyridine *N*-oxide contains a good leaving group at position C-3 (Br, OMe), the 2,3-dihydropyridine intermediate undergoes facile elimination to give a fused pyridine oxazolone ring system. Further studies revealed that the highest conversions of the starting *N*-oxide and yields of cycloadducts are obtained in polar solvents (DMF, DMSO, dioxane), which suggests a polar or stepwise mechanism for this cycloaddition, in accordance with theoretical calculations results discussed in Sect. 2.

Concerning cycloaddition of azoles *N*-oxides with isocyanates, reactions of 1-methylbenzimidazole 3-oxide with phenyl isocyanate, phenyl isothiocyanate, or even benzonitrile give 2-substituted products of the opening of the initial five-membered cycloadduct (Scheme 23) [41, 49]. The reaction of 1-benzyl-4,5-dimethylimidazole 3-oxide with PhNCO has been reported to give a *N*-imidazolyl-*N,N′*-diphenylurea derivative, a product of initial cycloaddition, formation of 2-aminoimidazole, and its subsequent reaction with the second molecule of PhNCO [48]. In 2000, these results were confirmed and investigated in greater detail by Mlostoń and coworkers, who reported that imidazole *N*-oxides of various substitution patterns react at rt with isocyanates and isothiocyanates to give 2-aminoimidazoles [51]. The reaction probably occurs via a stepwise mechanism since both isocyanates and isothiocyanates give the same products, although in typical concerted dipolar cycloaddition reactions the former react at the C=N bond and the latter at the C=S bond.

An interesting example is cycloaddition of pyridine *N*-oxides with 2- and 4-pyridyl isocyanates derivatives, generated in situ by a Curtius rearrangement of the corresponding acyl azides, that in turn had been prepared form carboxylic acids upon treatment with diphenylphosphoryl azide (Scheme 24) [103, 104]. These reactions, performed in dry boiling benzene, yield bis(pyridyl)amines in 65–80% yield. In the case of 3-nitro-4-pyridyl isocyanate, an intermediate resulting from

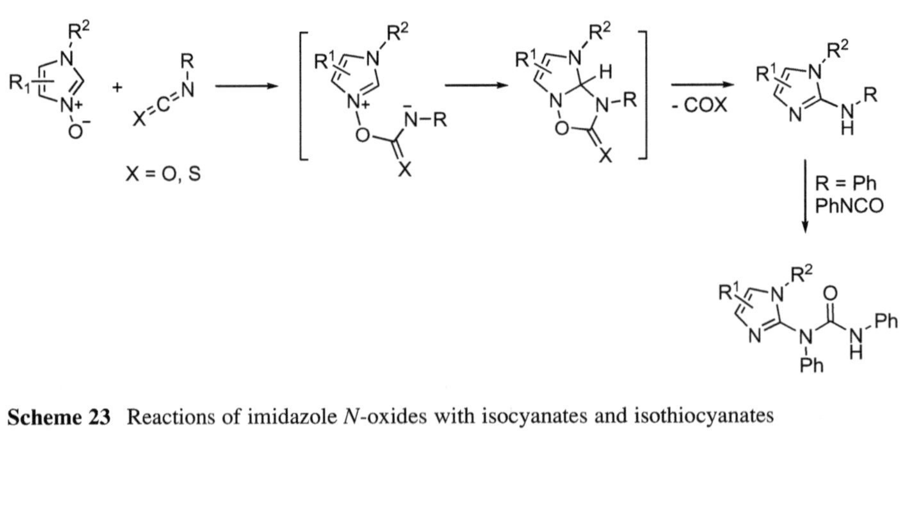

Scheme 23 Reactions of imidazole *N*-oxides with isocyanates and isothiocyanates

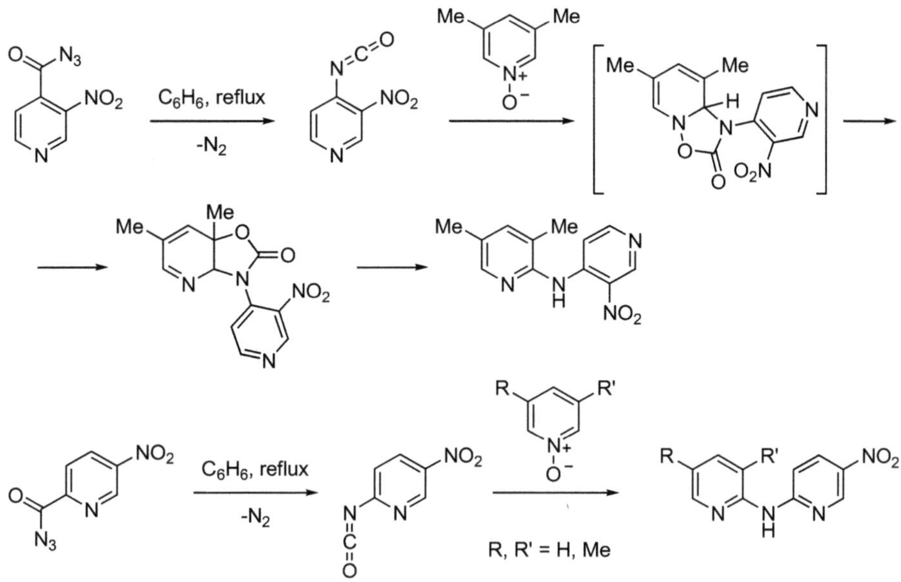

Scheme 24 Dipolar cycloaddition of azine *N*-oxides with heteroaryl isocyanates generated in situ via Curtius rearrangement

1,5-sigmatropic rearrangement of the initial cycloadduct could be detected in the crude reaction mixture.

2-Heteroarylisocyanates generated in situ from the respective 2-carboxylic acids have been employed in cycloaddition with imidazole *N*-oxides leading to bis (heteroaryl)amines with chelating properties [105]. Coordination of these amines to boron trifluoride provided analogues of aza-BODIPY fluorescent dyes, characterized by high thermal and photochemical stability and very large Stokes shift values, resulting in strong fluorescence also in crystalline phase (Scheme 25).

A considerable research effort has been directed toward development of 2-amidation of azines via cycloaddition of *N*-oxides with various forms of highly electrophilic nitrilium cations. Already in 1970s, Abramovitch and coworkers investigated reactions of *N*-oxides with imidoyl chlorides and nitrilium salts prepared in advance. These reactions are believed to proceed as stepwise cycloadditions and lead to the expected 2-acylamidoazines, together with side products resulting from a 1,5-sigmatropic rearrangement [106].

Much later, Manley and Bilodeau reported that imidoyl chlorides could be generated in situ from secondary amides upon the action of oxalyl chloride (or phosgene in the case of acetanilide) [107]. Their reactions with pyridine, quinoline, and isoquinoline *N*-oxides produce 2-aminopyridine amides. This protocol is mild and selective compared to the reactions with isolated imidoyl chlorides, as, for example, 2-methylpyridine reacts only at the position 6 of the ring and not at the methyl group.

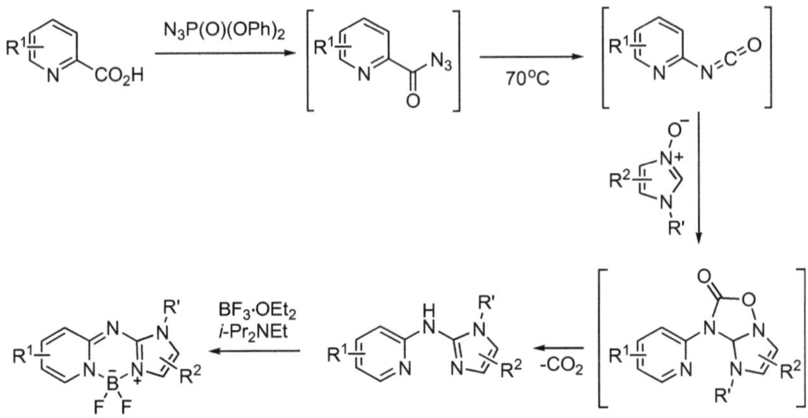

Scheme 25 Application of dipolar cycloaddition of heteroaryl isocyanates and imidazole *N*-oxides for the preparation of unsymmetrical bis(heteroaryl)amine ligands

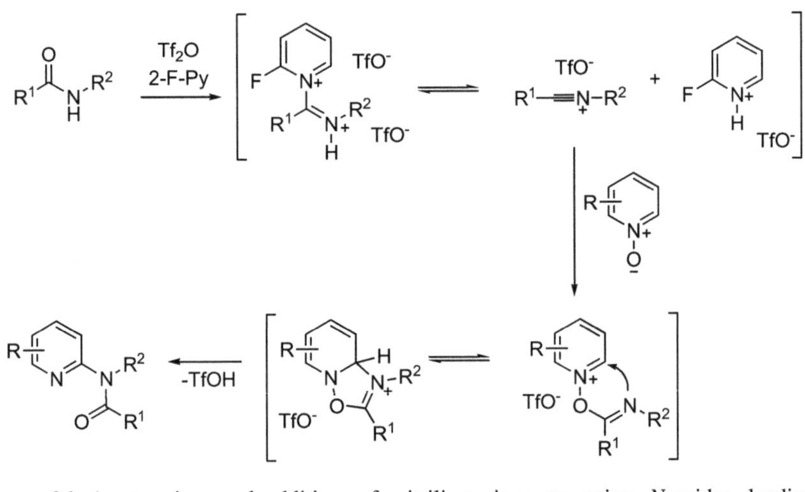

Scheme 26 A stepwise cycloaddition of nitrilium ions to azine *N*-oxides leading to 2-acylaminoazines

In a similar reaction, pyridinylated, isoquinolinylated, and quinolinylated amides are formed in the reaction of *N*-oxides with secondary amides activated with trifluoromethanesulfonic anhydride (Tf_2O) in the presence of base [108]. 2-Fluoropyridine, or in some cases 2-chloropyridine, was found to be the base of optimum nucleophilicity. Detailed mechanistic studies showed that varying amounts of a nitrilium ion are formed, which is in equilibrium with an amidinium ion. From the experiments with deuterated *N*-oxides, the authors concluded that an isoxazoline intermediate is formed in an irreversible nucleophilic addition of *N*-oxide to the nitrilium ion, followed by a reversible nucleophilic attack of the nitrogen atom on the heterocyclic ring (Scheme 26).

N-(2-Heteroaryl)amides can be efficiently obtained in a reaction between azine *N*-oxides and isocyanides activated by trimethylsilyl triflate (TMSOTf) [77]. The reaction is probably a stepwise cycloaddition of *N*-oxide with a nitrilium cation generated from isocyanide and TMSOTf. The main indication for such a mechanism is the regioselectivity of the reactions with pyridine *N*-oxides containing various groups at the 3 position and the interplay between steric and electronic factors: the presence of 3-alkyl groups promotes the formation of the 2,5-substituted products, whereas the presence of electron-withdrawing substituents which increase positive charge at the neighboring 2 position leads to increased amounts of the 2,3-isomers.

An interesting development has been described by Couturier and coworkers, who reported that acyloyl isocyanates can be generated upon the action of oxalyl chloride on primary amides in the absence of base [109]. In the presence of base, a facile transformation of amide to nitrile occurs. Cycloaddition of thus prepared acyloyl isocyanates with *N*-oxides of quinolines, isoquinolines, and phenanthridines allows preparation of *N*-acylated 2-aminoazines from primary amides.

5 Cycloaddition to Carbon–Sulfur Double Bonds

An isolated example of a reaction between pyridine *N*-oxide and a sulfin has been described, namely, bis(trifluoromethyl)sulfin, which smoothly produced 2-hexafluoroisopropylpyridine [110]. This outcome can be most likely rationalized by cycloaddition of *N*-oxide and the C=S double bond of sulfin followed by the N–O bond cleavage and loss of SO_2 (Scheme 27).

Cycloaddition to thioketones has been studied extensively by the Mlostoń group [111–113]. The reaction of imidazole *N*-oxides with 2,2,4,4-tetramethyl-3-thioxocyclobutanone and other thioketones results in the formation of imidazole-2(3*H*)-thiones and replacement of sulfur for oxygen in the initial thioketone. In contrast, azine *N*-oxides fail to react with thioketones. These results can be explained by ring opening of the intermediate 1,4,2-oxathiazolidine cycloadduct with restoration of aromaticity of imidazole ring. Cycloaddition is probably stepwise since thiofluorenone, which is a very good dipolarophile but particularly poor sulfur nucleophile, does not react. A stable thioketene reacts with imidazole *N*-oxides via a similar zwitterionic intermediate, but the dominating pathway is departure of imidazole as a leaving group associated with thiirane ring closure.

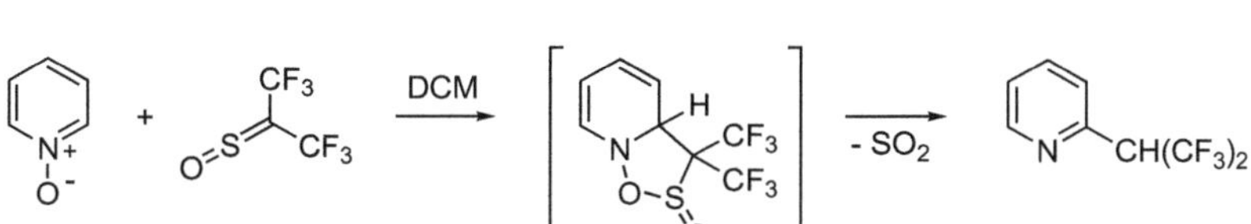

Scheme 27 Cycloaddition of pyridine *N*-oxide and bis(trifluoromethyl)sulfin

6 Conclusions

Over the years, 1,3-dipolar cycloaddition of aromatic *N*-oxides has evolved from a set of isolated examples of sometimes complex, bizarre reactions to an important field of heterocyclic chemistry. The success of the new *N*-oxide cycloaddition reactions developed in the recent years relays upon predictable and controllable formation and further transformations of the intermediate isoxazoline or isoxazolidine cycloadducts. The discovery of such reactions has provided useful synthetic methods of selective introduction of complex substituents into the heteroaromatic ring (including valuable fluorinated derivatives) in the position originally occupied by hydrogen, under close to neutral conditions (no strong acids nor bases) and without the need for transition metal catalysts. Exploration of reactions with new dipolarophiles will certainly bring further developments in this area of chemistry.

References

1. Youssif S (2001) ARKIVOC 2001(1):242
2. Begtrup M (2012) Adv Heterocycl Chem 106:1
3. Liu C, Luo J, Xu L, Huo Z (2013) ARKIVOC 2013(1):154
4. Wang Y, Zhang L (2015) Synthesis 47:289
5. Mlostoń G, Jasiński M, Wróblewska A, Heimgartner H (2016) Curr Org Chem 20:1359
6. Bull JA, Mousseau JJ, Pelletier G, Charette AB (2012) Chem Rev 112:2642
7. Andersson H, Olsson R, Almqvist F (2011) Org Biomol Chem 9:337
8. Larionov OV, Stephens D, Mfuh A, Chavez G (2014) Org Lett 16:864
9. Nishida T, Ida H, Kuninobu Y, Kanai M (2014) Nat Commun 5:3387
10. Bering L, Antonchick AP (2015) Org Lett 17:3134
11. Zhang Z, Pi C, Tong H, Cui X, Wu Y (2017) Org Lett 19:440
12. Bugaenko DI, Yurovskaya MA, Karchava AV (2017) J Org Chem 82:2136
13. Mąkosza M, Wojciechowski K (2004) Chem Rev 104:2631
14. Iwasaki G, Wada K, Saeki S, Hamana M (1984) Heterocycles 22:1811
15. Zhang S, Liao L-Y, Zhang F, Duan X-F (2013) J Org Chem 78:2720
16. Schlosser M, Mongin F (2007) Chem Soc Rev 36:1161
17. Kanyiva KS, Nakao Y, Hiyama T (2007) Angew Chem Int Ed 46:8872
18. Cho SH, Hwang SJ, Chang S (2008) J Am Chem Soc 130:9254
19. Wu J, Cui X, Chen L, Jiang G, Wu Y (2009) J Am Chem Soc 131:13888
20. Xiao B, Liu Z-J, Liu L, Fu Y (2013) J Am Chem Soc 135:616
21. Li G, Jia C, Sun K (2013) Org Lett 15:5198
22. Chen X, Zhu C, Cui X, Wu Y (2013) Chem Commun 49:6900
23. Sharma U, Park Y, Chang S (2014) J Org Chem 79:9899
24. Kianmehr E, Faghih N, Khan KM (2015) Org Lett 17:414
25. Sun W, Wang M, Zhang Y, Wang L (2015) Org Lett 17:426
26. Odani R, Hirano K, Satoh T, Miura M (2015) J Org Chem 80:2384
27. Neufeldt SR, Jiménez-Osés G, Huckins JR, Thiel OR, Houk KN (2015) J Am Chem Soc 137:9843
28. Liu J-B, Sheng X-H, Sun C-Z, Huang F, Chen D-Z (2016) ACS Catal 6:2452
29. Zhang W-M, Dai J-J, Xu J, Xu H-J (2017) J Org Chem 82:2059

30. Campeau L-C, Bertrand-Laperle M, Leclerc J-P, Villemure E, Gorelsky S, Fagnou K (2008) J Am Chem Soc 130:3276
31. Liégault B, Lapointe D, Caron L, Vlassova A, Fagnou K (2009) J Org Chem 74:1826
32. Liu W, Li Y, Wang Y, Kuang C (2013) Eur J Org Chem 2013:5272
33. Zhu C-L, Zhang Y-Q, Yuan Y-A, Xu H (2015) Synlett 26:345
34. Ryzhakov AV, Rodina LL (1992) Khim Geterotsikl Soedin 1992:579; (1993) Chem Abstr 118:168895
35. Sustmann R (1974) Pure Appl Chem 40:569
36. Matsuoka T, Harano K (1995) Tetrahedron 51:6451
37. Harano K, Kondo R, Murase M, Matsuoka T, Hisano T (1986) Chem Pharm Bull 34:966
38. Hisano T, Harano K, Matsuoka T, Suzuki T, Murayama Y (1990) Chem Pharm Bull 38:605
39. Matsuoka T, Harano K, Hisano T (1994) Heterocycles 37:257
40. Ishiguro Y, Yoshida M, Funakoshi K, Saeki S, Hamana M (1983) Heterocycles 20:193
41. Takahashi S, Kanō H (1965) J Org Chem 30:1118
42. Abramovitch RA, Grins G, Rogers RB, Shinkai I (1976) J Am Chem Soc 98:5671
43. Abramovitch RA, Shinkai I (1976) Acc Chem Res 9:192
44. Nour El-Din AM, Tawik A-RM, Ramadan M (1988) Rev Roum Chim 33:183
45. D'Errico S, Piccialli V, Oliviero G, Borbone N, Amato J, D'Atri V, Piccialli G (2011) Tetrahedron 67:6138
46. Raminelli C, Liu Z, Larock RC (2006) J Org Chem 71:4689
47. Shaibu BS, Kawade RK, Liu R-S (2012) Org Biomol Chem 10:6834
48. Ferguson IJ, Schofield KJ (1975) Chem Soc Perkin Trans 1:275
49. Takahashi S, Kanō H (1964) Chem Pharm Bull 12:1290
50. Jones RCF, Martin JN, Smith P, Gelbrich T, Light ME, Hursthouse MB (2000) Chem Commun 2002:1949
51. Mlostoń G, Gendek T, Heimgartner H (2000) Tetrahedron 56:5405
52. Mlostoń G, Jasiński M, Heimgartner H (2011) Eur J Org Chem 2011:2542
53. Huisgen R, Seidl H (1963) Tetrahedron Lett 4:2023
54. Huisgen R, Seidl H, Wulff J (1969) Chem Ber 102:915
55. Hamana M, Funakoshi K, Shigyo H, Kuchino Y (1975) Chem Pharm Bull 23:346
56. Ding Q, Wang D, Luo P, Liu M, Pu S, Zhou L, Beilstein J (2013) Org Chem 9:1949
57. Ye S, Gao K, Wu J (2010) Adv Synth Catal 352:1746
58. Kumar R, Kumar I, Sharma R, Sharma U (2016) Org Biomol Chem 14:2613
59. Xia H, Liu Y, Zhao P, Gou S, Wang J (2016) Org Lett 18:1796
60. Mlostoń G, Urbaniak K, Wojciechowska A, Linden A, Heimgartner H (2012) Helv Chim Acta 95:577
61. Mlostoń G, Jasiński M, Linden A, Heimgartner H (2006) Helv Chim Acta 89:1304
62. Hisano T, Harano K, Matsuoka T, Matsuzaki T, Eto M (1991) Chem Pharm Bull 39:537
63. Heaney F, Lawless E, Mahon M, McArdle P, Cunningham D (2006) Org Biomol Chem 4:2408
64. An Y, Zheng D, Wu J (2014) Chem Commun 50:9165
65. Lam Y, Stanway SJ, Gouverneur V (2009) Tetrahedron 65:9905
66. Nguyen TB, Martel A, Gaulon-Nourry C, Dhal R, Dujardin G (2012) Org Prep Proced Int 44:1
67. Liu JT, Jin QH, Lü HJ, Huang WY (2001) Tetrahedron Lett 42:5937
68. Tomashenko OA, Grushin VV (2011) Chem Rev 111:4475
69. Studer A (2012) Angew Chem Int Ed 51:8950
70. Wu X-F, Neumann H, Beller M (2012) Chem Asian J 7:1744
71. Besset T, Schneider C, Cahard D (2012) Angew Chem Int Ed 51:5048
72. Wang J, Sánchez-Roselló M, Aceña JL, del Pozo C, Sorochinsky AE, Fustero S, Soloshonok VA, Liu H (2014) Chem Rev 114:2432
73. Lantaño B, Torviso MR, Bonesi SM, Barata-Vallejo S, Postigo A (2015) Coord Chem Rev 285:76

74. Mailey EA, Ocone LR (1968) J Org Chem 33:3343
75. Banks RE, Haszeldine RN, Robinson JM (1976) J Chem Soc, Perkin Trans 1 1976:1226
76. Loska R, Mąkosza M (2006) Mendeleev Commun 16:161
77. Vamos M, Cosford NDP (2014) J Org Chem 79:2274
78. Ichikawa J, Wada Y, Fujiwara M, Sakoda K (2002) Synthesis 2002:1917
79. Loska R, Mąkosza M (2008) Chem Eur J 14:2577
80. Fuqua SA, Duncan WG, Silverstein RM (1965) J Org Chem 30:1027
81. Hayashi S, Nakai T, Ishikawa N, Burton DJ, Naae DG, Kesling HS (1979) Chem Lett 8:983
82. Zheng J, Cai J, Lin J-H, Guo Y, Xiao J-C (2013) Chem Commun 49:7513
83. Qiao Y, Si T, Yang M-H, Altman RA (2014) J Org Chem 79:7122
84. Zhao Y, Huang W, Zhu L, Hu J (2010) Org Lett 12:1444
85. Zhu L, Li Y, Zhao Y, Hu J (2010) Tetrahedron Lett 51:6150
86. Wang X-P, Lin J-H, Xiao J-C, Zheng X (2014) Eur J Org Chem 2014:928
87. Raghavanpillai A, Burton DJ (2006) J Org Chem 71:194
88. Gøgsig TM, Søbjerg LS, Lindhardt (neé Hansen) AT, Jensen KL, Skrydstrup TJ (2008) Org Chem 73:3404
89. Fujita T, Ichitsuka T, Fuchbe K, Ichikawa J (2011) Chem Lett 40:986
90. Yamamoto T, Yamakawa T (2012) Org Lett 14:3454
91. Duric S, Schmidt BM, Ninnemann NM, Lentz D, Tzschucke CC (2012) Chem Eur J 18:437
92. Ichitsuka T, Takanohashi T, Fujita T, Ichikawa J (2015) J Fluor Chem 170:29
93. Chelucci G (2012) Chem Rev 112:1344
94. Loska R, Szachowicz K, Szydlik D (2013) Org Lett 15:5706
95. Loska R, Bukowska P (2015) Org Biomol Chem 13:9872
96. Szpunar M, Loska R (2015) Eur J Org Chem 2015:2133
97. Seidl H, Huisgen R, Grashey R (1969) Chem Ber 102:926
98. Hisano T, Yoshikawa S, Muraoka K (1974) Chem Pharm Bull 22:1611
99. Hisano T, Matsuoka T, Ichikawa M (1976) Chem Pharm Bull 24:533
100. Hisano T, Matsuoka T, Tsutsumi K, Muraoka K, Ichikawa M (1981) Chem Pharm Bull 29:3706
101. Hisano T, Matsuoka T, Fukunaga K, Ichikawa M (1982) Chem Pharm Bull 30:3776
102. Matsuoka T, Harano K, Kubo H, Hisano T (1986) Chem Pharm Bull 34:572
103. Holt J, Fiksdahl A (2007) J Heterocycl Chem 44:375
104. Holt J, Andreassen T, Bakke JM, Fiksdahl A (2005) J Heterocycl Chem 42:259
105. Bukowska P, Piechowska J, Loska R (2017) Dyes Pigments 137:312
106. Abramovitch RA, Rogers RB, Singer GM (1975) J Org Chem 40:41
107. Manley PJ, Bilodeau MT (2002) Org Lett 4:3127
108. Medley JW, Movassaghi M (2009) J Org Chem 74:1341
109. Couturier M, Caron L, Tumidajski S, Jones K, White TD (2006) Org Lett 8:1929
110. Schwab M, Sundermeyer W (1988) Chem Ber 121:75
111. Mlostoń G, Gendek T, Heimgartner H (1998) Helv Chim Acta 81:1585
112. Pieczonka AM, Mlostoń G, Heimgartner H (2012) Helv Chim Acta 95:404
113. Wróblewska A, Mlostoń G, Heimgartner H (2015) Tetrahedron Asymmetry 26:505

Top Heterocycl Chem (2017) 53: 111–152
DOI: 10.1007/7081_2017_4

Published online: 5 May 2017

Recent Advances in the Photochemistry of Heterocyclic *N*-Oxides and Their Derivatives

James S. Poole

Abstract The photochemistry of the heteroarene *N*-oxides and their derivatives is complex and multifaceted. This review discusses the photochemistry of heteroarene *N*-oxides in light of new studies of these species, as well as recent studies of nitrene intermediates that have been proposed as potential photochemical intermediates. Recent developments in the photochemistry of *N*-alkoxyheteroarenium salts are also discussed, as well as the photochemistry of the *N*-hydroxypyridone and *N*-hydroxypyrithione systems, which may be thought of as tautomeric forms of the *N*-oxides.

Keywords Acyloxy radicals • Barton esters • Carbon-centered free radicals • Cationic polymerization • Electron transfer • 2*H*-Azirines • Hydroxyl radical Alkoxyl radicals • Isoquinoline *N*-Oxides • *N*-Alkoxypyridinium salts • *N*-Hydroxy-2-pyridone • *N*-Hydroxypyridine-2-thione • Nitrile ylides • 1,3-Oxazepines • Photochemistry • Photopolymerization • Pyridine *N*-Oxides • Quinoline *N*-Oxides • Radical polymerization • Vinyl nitrenes

Contents

J.S. Poole (✉)
Department of Chemistry, Ball State University, Muncie, IN 47306, USA
e-mail: jspoole@bsu.edu

1 Introduction

The photochemistry of the aza-aromatic *N*-oxides is a field rich with chemistry, and has excited a great deal of interest in the organic community. There have been a number of reviews in the past on this subject, including those of Spence et al. in 1970 [1], Bellamy and Streith, and LaBlache-Combier (both in 1976) [2, 3], Albini and Alpegiani in 1984 [4], Albini et al. in1995 [5], and Albini and Fagnoni in 2004 [6]. This contribution proposes to describe recent advances in this field, including the photochemistry of closely allied structures. Whilst a certain degree of retrospection is unavoidable, this contribution does not seek to replicate the aforementioned works, and the reader is encouraged to seek out these contributions to provide a broader perspective of the field in the appropriate historical context.

In this review, recent studies in the photochemistry of heteroarene *N*-oxides (in which the compound pyridine *N*-oxide **1** acts as an archetype) are discussed. In addition, tautomeric forms of pyridine *N*-oxides such as *N*-hydroxypyridone (**2**), and pyrithione (**3**) and their derivatives, as well as systems such as *N*-alkoxylpyridinium salts (**4**) are considered.

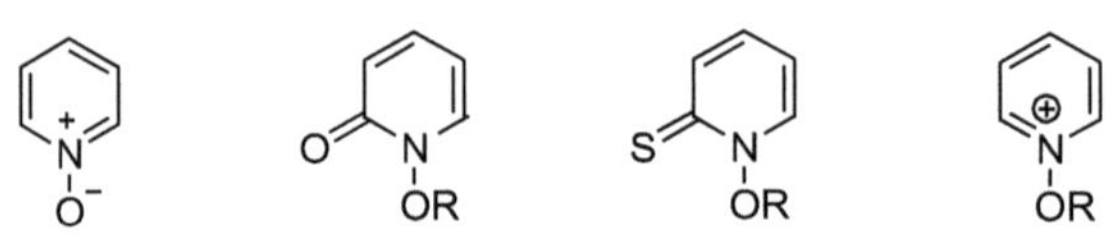

(**1**) (**2**) (**3**) (**4**)

This review includes discussion of experimental techniques such as matrix isolated infra-red spectroscopy and laser flash photolysis (LFP) that may not be familiar to the synthetic organic chemist. The reader is referred to the contributions of Dunkin [7] and Bally [8], for an introductory treatment of the matrix isolation experiment. An analogous contribution describing the LFP experiment is provided by Scaiano [9].

2 The Photochemistry of Aza-Arene *N*-Oxides

2.1 Pyridine N*-Oxides*

Pyridine *N*-oxide **1** represents the archetype of these species, and as such, its photochemistry has been studied under a range of conditions. A certain

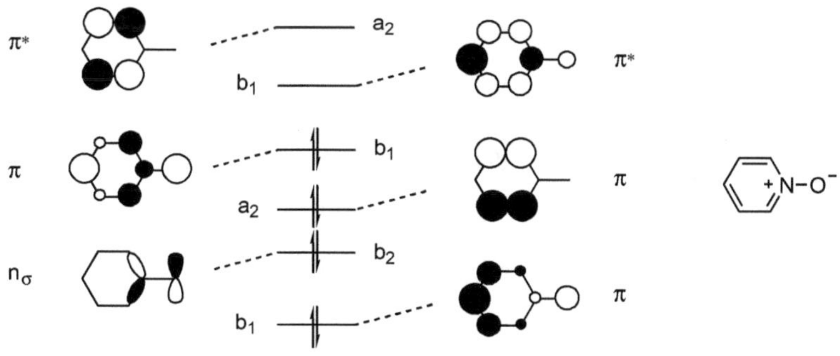

Scheme 1 Truncated qualitative MO diagram for pyridine *N*-oxide, HF/cc-pVTZ (energies not to scale)

amount of care must be exercised when describing the observed photochemistry of this species, since in many cases the photochemistry is dominated by polymerization, and isolated yields of small molecule photoproducts are low.

It has been established that photoexcitation of pyridine *N*-oxide in both gas [10, 11] and condensed phases [12] occurs via a $\pi \rightarrow \pi^*$ transition to the singlet excited state. The nature of the photoexcited states of pyridine *N*-oxide has been studied by a range of gas phase and computational techniques [13–16], most recently by Palmer et al. [17]. The molecular orbitals involved in the lower excited states are shown in Scheme 1.

MCSCF calculations indicate that there are two lowest energy transitions with non-zero oscillator strengths in the UV-Visible spectrum: $\pi(b_1) \rightarrow \pi^*(a_2)$ (3.65 eV) and $\pi(b_1) \rightarrow \pi^*(b_1)$ (4.61 eV). In either case, the excited state singlet will involve depopulation of electron density at O, and so the S_1 excited state (**1***) may be written with a degree of biradical character.

In the absence of triplet sensitization, the bulk of pyridine *N*-oxide photochemistry is believed to proceed via the singlet state. Recently, Albini et al. [5, 6] have cautioned against the notion that all observed photochemistry may be rationalized via a single mechanism.

2.1.1 Ring Contraction

The singlet photochemistry of the *N*-oxide systems has often been described as involving an oxaziridine intermediate (**5**) in a manner analogous to the photochemistry of nitrones. Albini et al. [5, 6] have argued against the mechanistic necessity of this structure as a discrete intermediate species given that it has yet to be experimentally observed – the structure **5**/**5**† may correspond to a transition state structure, or a structure associated with a conical intersection of closed-shell and biradical singlet states (Scheme 2).

Fragmentation of the biradical **6** leads to the formation of vinyl nitrene **7**. Singlet vinyl nitrenes are highly reactive intermediates that have yet to be observed

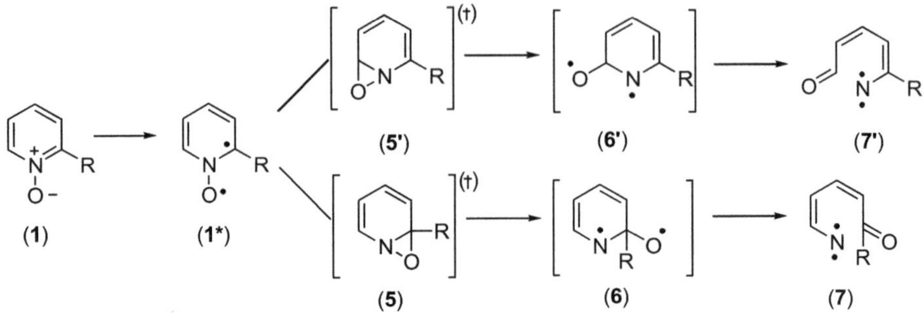

Scheme 2 Formation of biradical intermediates in the singlet photochemistry of a substituted pyridine *N*-oxide

experimentally. Calculations at the CASSCF level of theory indicate that the lowest singlet state is a so-called open shell singlet [18], indicating that **10**, if formed, would exhibit biradical behavior (Scheme 3), and would be expected to rapidly cyclize over a very small barrier to yield the 2*H*-azirine **8**. The lifetime of the singlet vinyl nitrene would be expected to be small: phenyl nitrene, which also has an open-shell configuration as its lowest energy singlet state [19], has a sub-nanosecond lifetime at room temperature [20].

The photochemistry of vinyl azides and nitrenes, and 2*H*-azirines has been a field of some considerable study in recent years [21]. In particular, 2*H*-azirines and vinyl nitrenes have been suggested as intermediates for the photochemical conversion of isoxazoles to oxazoles [22–24]. Matrix isolation studies by Fausto et al. suggest that isomerization proceeds as described in Scheme 4 [25].

Gudmundsdottir et al. have demonstrated the ability to generate and detect triplet vinyl nitrenes in solution (and in argon matrices) by internal sensitization [26]. These species are generated by the photolysis of 2*H*-azirines or azides containing groups which enhance intersystem crossing in the azirine excited state (for example, **10** and **11**). These species are relatively long-lived, and exhibit typical triplet behavior, in that they undergo dimerization, add to alkenes, and are trapped by oxygen, leading to nitrile species (although it may be argued in the latter case that oxygen attack on the triplet azide precursor occurs simultaneously with nitrogen extrusion) [27].

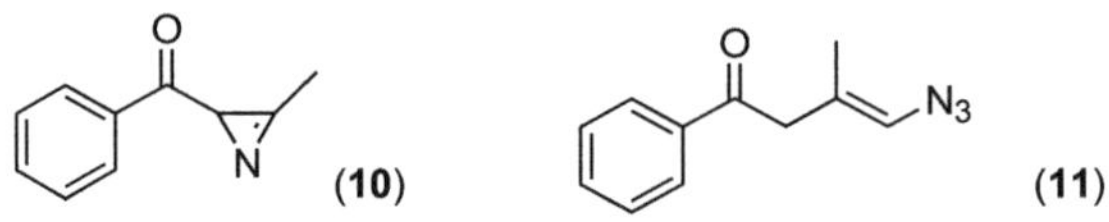

One of the most commonly observed photochemical products associated with pyridine *N*-oxide systems are 2-acylpyrrole species, arising from ring contraction reactions (Scheme 3). There are a number of examples in the literature where a nominal vinyl nitrene may be trapped with a suitably emplaced double bond to yield rings [28, 29], and analogous cyclizations may yield the acylpyrrole **9**. The trapping of

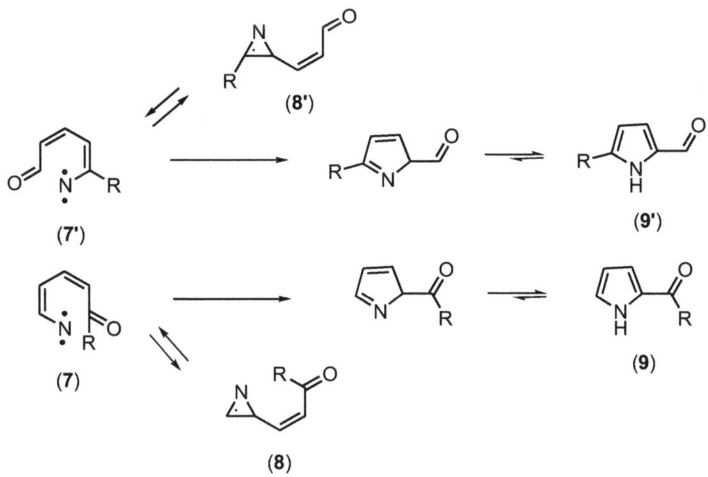

Scheme 3 The chemistry of singlet vinyl nitrenes

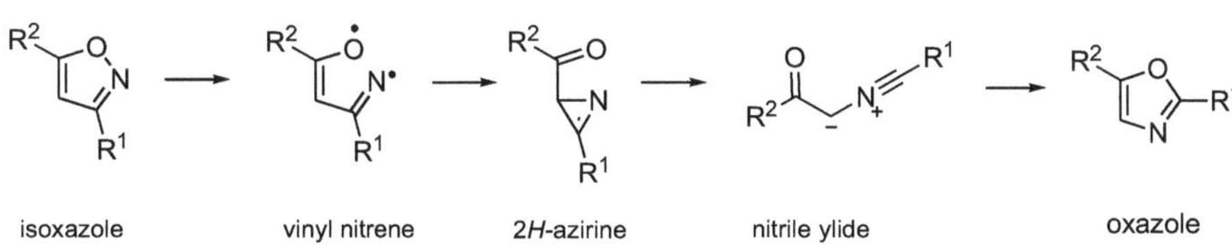

Scheme 4 Isomerization of isoxazoles to oxazoles in argon matrices

vinyl nitrenes has been observed in both intra- and inter-molecular modes [30]. It has been noted that the isolated yield of **9** in aqueous solution is greatly enhanced by the addition of Cu^{II} salts [31], and that the intermediacy of Cu^{I} species was indicated since yields were markedly reduced under oxidizing conditions [32]. Recently Chen et al. have reported that the thermal reaction of vinyl azides with aryl alcohols to generate disubstituted pyrroles yielded different regioselectivities when using Cu^{II} and Ni^{II} salts [33]. Their proposed mechanism involved formal entrapment of the nitrene as an enaminyl radical by electron transfer from Cu^{I}, although it is not clear whether this occurs in conjunction with, or subsequent to, deazotization of the azide.

Small amounts of other ring-contracted species have been observed in the photochemistry of *N*-oxides: Alkaitas and Calvin observed the formation of *N*-formylpyrrole in the photochemistry of pyridine *N*-oxide in alcohol (2–4%) [34]; and Bellamy et al. have reported a number of pyrrole species in the photochemistry of 2- and 3-substituted pyridine *N*-oxide systems in aqueous solution, again in small yields (Scheme 5) [35].

The proposed mechanisms for the formation of these species involve potential 1,5- (and subsequent 1,3-) sigmatropic shifts in the pyrrolenine intermediate **12**. An

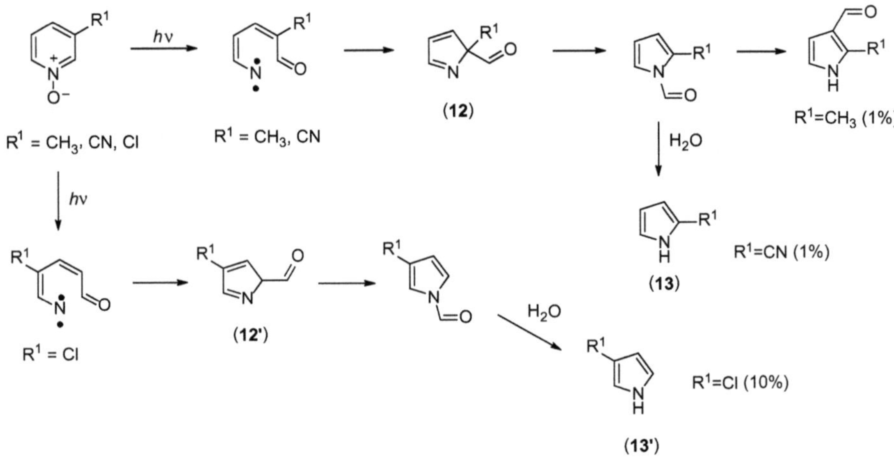

Scheme 5 Aqueous photochemistry of 3-substituted pyridine 1-oxides

alternative mechanism for the formation of the chloroderivative involves the hydrolysis of a 1,3-oxazepine intermediate similar to **14** (see below). The *N*-formyl species are susceptible to further hydrolysis in solution or during isolation to yield the pyrroles **13** and **13**′.

2.1.2 Ring Expansion

There are a number of possible modes of ring expansion available to the simple pyridine *N*-oxides that lead to 1,2-oxazepines **15**, and the 1,3-oxazepines **14**. Neither species has been reported for pyridine *N*-oxide, but it is worth noting that these species corresponding to **14** have been reported for a range of tri-, tetra-, and penta-aryl substituted species (i.e., systems analogous to those favoring formation of the ylide) in reasonable isolated (71–87%) yield [36], and in more modest yields for 2,6-dicyanopyridine N-oxides [37]. The 1,2-oxazepine **15** may in principle be formed directly from the excited state of pyridine *N*-oxide (Scheme 6). Fragmentation of the N–O bond would yield the vinyl nitrene **7**.

In principle, such a reaction might be reversible. However, studies on aryl nitrenes indicate that singlet phenyl nitrenes may be trapped by ketones and esters sited *ortho-* to the nitrene center [38, 39]. Substitution of an α, β-unsaturated carbonyl moiety will lead to cyclization to form indole derivatives, rather than 1,2-benzoxaxepines [40].

1,3-Oxazepines may form from the biradical intermediate **6** via structures such as **16** that may be either an intermediate or a transition state (Scheme 7). Alternatively, the photochemistry of 2*H*-azirines may proceed via N–C bond scission to yield nitrene **7**, or C–C scission to yield the nitrile ylides **17**. Cyclization to yield structures such as **14** may then proceed in a manner similar to the formation of oxazoles in Scheme 4.

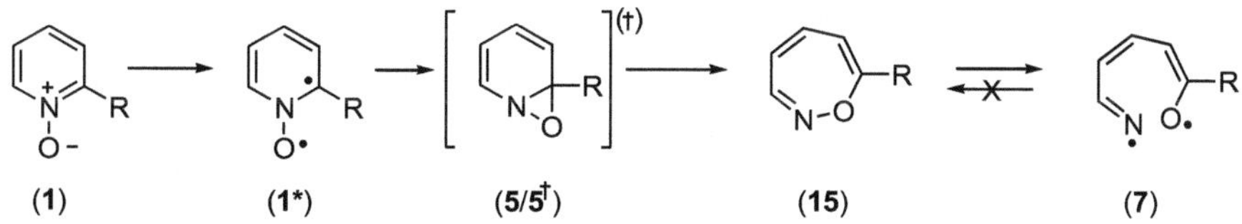

Scheme 6 Potential formation of 1,2-oxazepine **15**

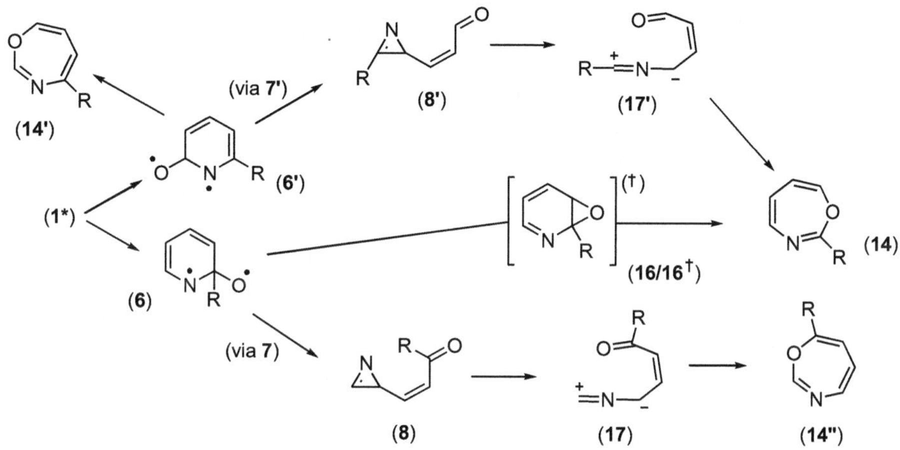

Scheme 7 Potential formation of 1,3-oxazepines **14, 14′, 14″**

The effects of substitution on the cleavage of 2*H*-azirine ring have been the subject of a significant amount of recent work, and the results are summarized in Table 1. In matrix isolated experiments, workers have also identified a ketenimine intermediate **21** that is generally accepted to arise from the triplet nitrene **20** by a 1,2-alkyl migration in the biradical (Scheme 8) [25]. There is some literature precedent for alkyl group migrations in 1,3-di-yl systems [41]. 1,2-hydrogen shifts in 1,3-diyl radicals have been postulated for the observed reactivity of 1,2-benzisoxazole and isoxazole in argon matrices [42]. In the case of 1,2-benzisoxazole, the ketenemine formed ($R^1 = H$), rearranges in the argon matrix to form *o*-hydroxybenzonitrile, a product reported in photolysis of 1,2-benzisoxazole in non-polar solutions [43].

Computational studies on these systems indicate that two modes of ring scission may operate, and the dominant mode is dependent on the substitution of the azirine ring (Table 1). Systems containing stabilizing groups at the 2-position (Case I) appear to favor the formation of the nitrile ylide, whereas systems with stabilizing groups at the 3-position (Case II) appear to increasingly favor the formation of the triplet nitrene species. The parent azirine is computed to favor the formation of the nitrile ylide, as is the system that contains stabilizing groups at both 2- and 3 positions (Case III). The photochemistry of simple pyridine 1-oxides would appear to generally correspond to

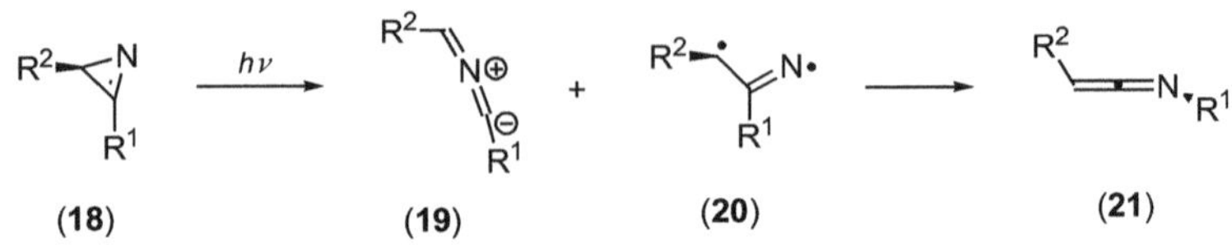

Scheme 8 Potential photochemistry of 2*H*-azirine intermediates

Table 1 Summary of observed photochemistry of 2*H*-azirines (Scheme 8)

18		Conditions	Nitrile ylide	Nitrene	Ketenimine
R^1	R^2		**19**	**20**	**21**
H	H	Computed [44]	Favored		
H	CHO	Ar matrix, 15 K [34]	Not observed	Observed[a]	Observed[a]
H	Ph	Computed		Favored	
CH_3	$COCH_3$	Ar matrix, 15 K [25]		Observed[a]	Observed[a]
CH_3	COPh	MeCN >334 nm [45]		Observed[b]	
		MeCN <300 nm	Observed		
CH_3	Ph	LFP in acetonitrile [46]		Observed	
		Ar matrix, 14 K			Observed
Ph	H	LFP in acetonitrile [47]	Observed		
		Computed	Favored		
(4-CN)Ph	H	LFP in acetonitrile	Observed		
(4-Cl)Ph	H	LFP in acetonitrile	Observed		
(4-CH_3)Ph	H	LFP in acetonitrile	Observed		
(4-OMe)Ph	H	LFP in acetonitrile	Observed		
(4-Ph)Ph	H	LFP in cyclohexane [48]	Observed[c]		
Ph	CH_3	LFP in acetonitrile	Observed		
		Ar matrix, 14 K	Observed		
Ph	Ph	LFP in acetonitrile	Observed		
Pyrenyl	H	Synthetic	Observed		

[a]Observed ketenimine formed via 1,2-hydrogen shift in nitrene
[b]Collapses to isoxazole, detectable by LFP
[c]Competes with intersystem crossing to yield triplet (non-reactive pathway)

Case I systems, whereas the polyphenyl substituted systems reported by Buchardt et al. correspond to Case II and Case III systems.

Diphenyl (**19**, $R^1 = R^2 =$ Ph), biphenyl (**19** $R^1 =$ 4′-biphenyl, $R^2 =$ H), and naphthyl ylides (**19** $R^1 = CH_3$, $R^2 =$ 1-naphthyl, also generated by the reaction of 1-naphthyl carbene with acetonitrile) [49], have lifetimes measured in the μs-ms range. The diphenyl and 4′-biphenyl ylides have been shown to be susceptible to protonation by alcohols, ultimately yielding the imine ethers [50]. In addition, ylides such as those in Table 1are known to react with a range of dipolarophiles [51], including unsaturated nitriles, esters, and amides [52, 53]. Similar photochemistry has been observed for 2-pyrenylazirines [51]. An analogous addition to aldehydes, believed to proceed via the azirinium radical cation generated by photoredox catalysis has also been reported [54, 55].

2.1.3 Ring Fragmentation

In basic aqueous solution, significant amounts of cyanoaldehydes **22**/**22′**or their enolates [56] may be generated, which may in turn be trapped with amines to generate cyanoenamines **23** [57–59]. There is evidence to suggest that the polymer formed upon photolysis contains cyano groups, and this may arise from the aldehyde derived from **22** (Scheme 9).

Formation of these species has been postulated to occur via proton transfer from a long-lived (63 ms lifetime, λ_{max}(EtOH) = 320 nm) vinyl nitrene [60], but this is inconsistent with what has been determined of the singlet chemistry of these species. Matrix and solution IR obtained following photolysis of pyridine *N*-oxide exhibited an absorbance peak at 2,145 cm^{-1} that the authors tentatively assigned to isocyanides [61]. A follow-up study indicated that the primary observed photoproduct had an absorbance at 2,188 cm^{-1}, and that the isocyanide is a secondary photoproduct (although it should be noted that for species such as **17** where $R_1 =$ H, proton transfer would yield an isocyanide) [62].

Studies on the thermolysis of vinyl azides, which would be expected to yield vinyl nitrenes indicate that for many systems where the vinyl azide is conjugated to a carbonyl group, nitrile products are obtained. Hassner et al. demonstrated that 2*H*-azirines could be thermally converted to nitriles [63]. The proposed mechanism for these rearrangements was held to be the fragmentation of the nitrene into an ion pair (a nitrile enolate and in the example shown, an acylium ion) which would then recombine to generate the nitrile (Scheme 10). In the case of rearrangements observed by the Moore group, the two moieties remained tethered [64, 65]. By extension, one may consider H^+ as potential leaving group in a similar mode, or alternatively, one may envisage a 1,2-hydride shift to generate the nitrile.

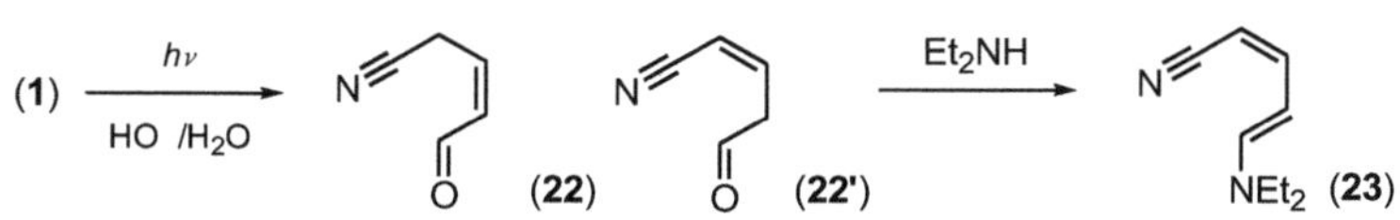

Scheme 9 Photochemistry of pyridine *N*-oxides in aqueous base

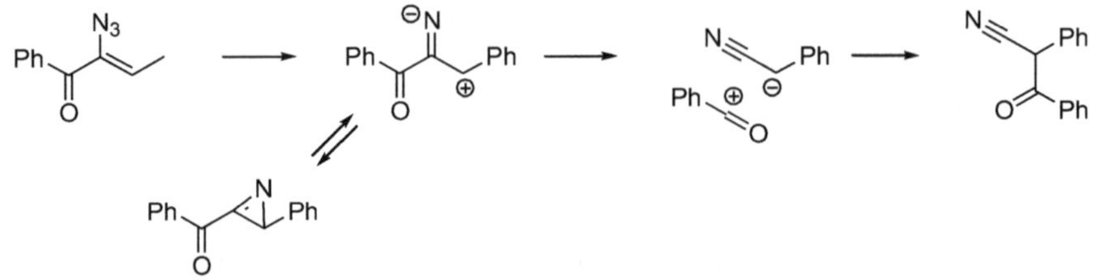

Scheme 10 Proposed mechanism for thermal formation of nitriles from azirines

However, the mechanism in Scheme 10 proceeds via a closed shell, zwitterionic intermediate that calculations suggest corresponds to the S_2 state of the vinyl nitrene, rather than to the more energetically favored S_1 and T_0 biradical states. This is unlikely to form thermally, and would be expected to be short-lived in a photochemical reaction, certainly with a half-life considerably shorter than the millisecond timescale. Alternatively, the S_1 or T_0 nitrenes (photolysis of a structurally similar azirine yields the triplet nitrene, see above) would require relatively unfamiliar 1,2-, 1,4-, or 1,6- hydrogen shifts in order to generate triplet biradicals that would collapse to form nitriles or ketenimines such as **21**. Deprotonation of **21** would lead to enolate anion precursors of **22/22′**. There is evidence for formal 1,2-hydrogen shift rearrangements in argon matrices (see Table 1) [43], but it is unclear how this would translate to solution phase chemistry.

Both the photolyses of 2*H*-azirines and the thermolyses of vinyl azides in Scheme 10 were carried out in non-polar or non-protic solvents under neutral conditions. Clearly, studies where possible, under aqueous or mixed solvent conditions, would be useful in this regard. Finally, an alternative pathway, involving hydroxide-mediated elimination from **15**, which has oxime ether characteristics, has literature precedent [66]. Oxime ethers are known to exhibit their own photochemistry to yield nitriles [67].

2.1.4 Deoxygenation

Photolysis of pyridine *N*-oxide is known to generate an oxidizing species that is capable of converting alkanes to alcohols [68], arenes to phenols [69], alcohols to aldehydes [33], alkenes to epoxides, allyl alcohols and ketones [70, 71], and thioethers to sulfoxides [72]. Similar reactions have been used to generate phosphine oxides from phosphine sulfides and selenides [73]. Copolymers containing pendant pyridine *N*-oxide groups have been assessed as potential photoresist materials – the crosslinking that occurs upon irradiation of these materials is believed to occur via the rupture of the N–O bonds [74]. Intramolecular oxidation has been observed to a small extent with 2-benzyl and 2-phenethylpyridine *N*-oxides to yield *o*′-hydroxy-substituted systems, and in the case of the former the benzylic alcohol [75]. Oxidations are known to be enhanced by the presence of Lewis acids such as

BF_3, the general principle being that complexation of the *N*-oxide reduces the ability for the excited state to rearrange on the singlet surface [76].

The energies of the triplet excited states of variously substituted pyridine *N*-oxides have been determined experimentally from their measured singlet-triplet absorption spectra – low intensity long-wavelength bands that are augmented under high pressure oxygen conditions. These triplet states are also of a $\pi \rightarrow \pi^*$ type [77, 78].

Bucher and Scaiano have reported the laser flash photolysis (LFP) of pyridine *N*-oxide in hydrocarbon solvents leads to the rapid formation of a long-lived (at least ms) product with λ_{max} of 335 nm [79]. In acetonitrile, these workers observed an overlying shorter-lived signal at 325 nm which they attributed to acetonitrile oxide (τ = 60 μs), formed by the reaction of $O(^3P)$ with acetonitrile. Stern-Volmer analysis of competitive experiments allowed the determination of rate coefficients for reaction of $O(^3P)$ with a range of organic substrates [80].

Albini et al. studied the photochemistry of 2- and 4-benzoylpyridine *N*-oxides as systems that would be strongly favored to exhibit triplet photochemistry. In the case of the 2-isomer, a mixture of deoxygenated starting material and *o*′-hydroxylated 2-benzoylpyridine was obtained, and in the case of the 4-isomer, only the deoxygenated product was obtained. In each case, the deoxygenation was considerably more efficient than most simple pyridine *N*-oxides [81]. Similarly, the deoxygenation reaction observed for 2-cyanopyridine was greatly enhanced by the presence of triplet sensitizers [82]. Thus, it would seem that the deoxygenation occurs via the triplet state. Carraher and Bakac have re-investigated the photochemistry of 4-benzoylpyridine *N*-oxide using LFP [83]. These workers found that the triplet state of the *N*-oxide was observable by LFP, but that the chemical yield of $O(^3P)$, determined by its reaction with cyclopentene, increased when the *N*-oxide was excited in the presence of triplet quenchers. This suggests that $O(^3P)$ was obtained from the excited singlet state.

2.1.5 Electron Transfer

Geletii et al. studied the potential photoinduced electron transfer between heteroarene *N*-oxides and bis-*N*-oxides to yield radical cation species [84]. These species may be observed to react with hydrocarbons under flash photolysis conditions. In the presence of oxygen, oxidation of hydrocarbons may be observed (the oxygen possibly required for hydrogen atom transfer). Thus, cyclohexane may be predominantly converted to cyclohexanone, with the alcohol as a minor product [85].

Pyridine *N*-oxides may also act as electron acceptors: Nakanishi et al. report the photoinduced electron transfer between the excited singlet state of 1-benzyl-1,4-dihydronicotinamide and pyridine *N*-oxide to generate the radical anion (Scheme 11) [86]. Protonation yields the azacyclohexadienyl radical which undergoes facile N–O bond cleavage to yield hydroxyl radical. Where X = H, the radical is calculated to exhibit significant pyramidalization, which enhances the fragmentation rate (see below). Where X = NO_2, the radical formed is planar, and generation of hydroxyl is not observed.

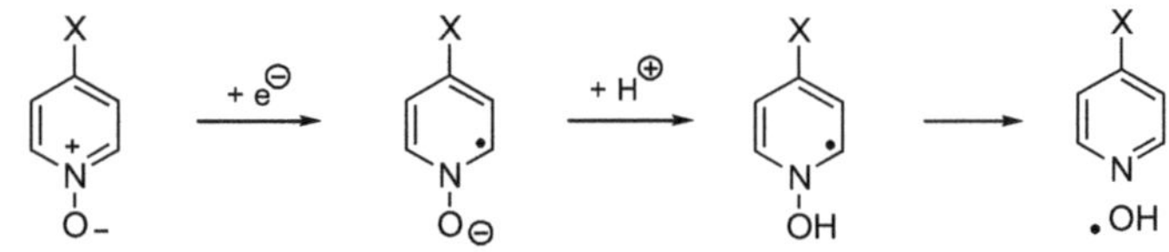

Scheme 11 Generation of hydroxyl radicals via photoinduced electron transfer

The ability of pyridine *N*-oxides to act as potential electron acceptors has been used to construct tethered donor-linker-acceptor species such as **24**, where the donor fragment consists of a dye capable of absorbing radiation in the visible part of the spectrum. These species have been evaluated as components of dye-sensitized solar cells, achieving efficiencies as high as 6.08% [87].

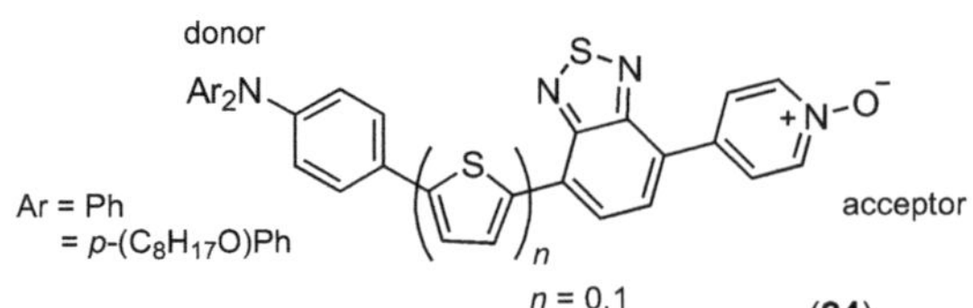

(**24**)

2.1.6 Other Monocyclic Mono *N*-Oxides

To a large extent, the photochemistry observed for other heteroaromatic systems is similar to that observed for pyridines, but there are some cases reported that are somewhat different. Pyridazine *N*-oxides undergo ring fragmentation to yield diazoketones, which in turn deazotize to generate cyclopropenyl ketones by singlet carbene chemistry (Scheme 12) [88]. Entrapment of the ketones with amines leads to the formation of *N*-alkylpyrroles [89], otherwise furans may be formed [90]. Alternatively, substitution at R^2 with phenyl, alkoxy or alkylamino groups yields dipolarophiles that may entrap the diazoalkane moiety to generate pyrazolines [91, 92]. These species are also capable of the more typical photooxygenation of alkenes [93].

2- and 5-methoxypyrimidine photochemistry yield outcomes that appear to be consistent with pyridine systems, although the presence of a methoxy group allows the formation of esters and carbamates upon fragmentation (Scheme 13) [94].

The pyrimidine chemistry is also observed in the purines. Thus, Bose et al. generated an isoguanosine derivative via photochemical fragmentation of formycin *N*(6)-oxide, possibly via the purinone [95]. Earlier studies on the simple purine *N*-oxides in water indicated the formation of purinone derivatives in a manner

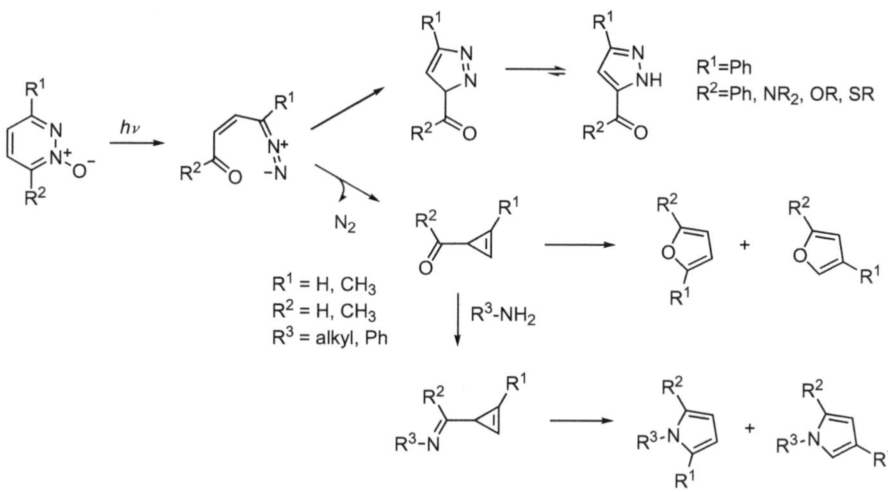

Scheme 12 Photochemistry of pyridazine *N*-oxides

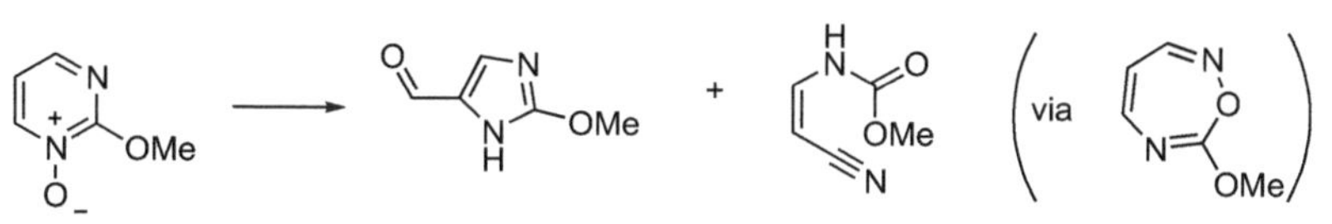

Scheme 13 Photochemistry of 2-methoxypyrimidine *N*-oxide

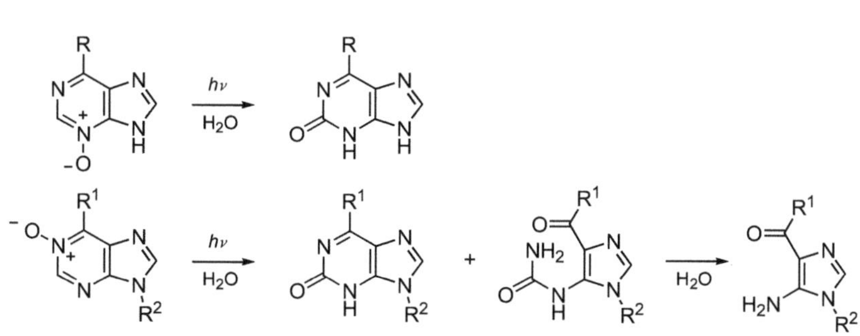

Scheme 14 Photochemistry of purine *N*-oxides and their derivatives

analogous to reactions observed for the quinoline *N*-oxide systems (Scheme 14 and see below) [96–100].

2.2 Bicyclic*N*-Oxides and Larger Systems

The photochemistry of quinoline *N*-oxide **25** is dissimilar in many respects to that of the pyridine species (Scheme 15, Table 2), including ring contraction to *N*-formylindole **27** in low yield (7%, with another 41% as a hydrolysis product) [101]. These indoles appear to correspond to hydrolysis products of **26**, and appear when reaction mixtures are subject to chromatography on silica gel [102]. Certain 1,3-oxazepines (**26**, $R^1 = CO_2Me$, $R^2 = H$, $R^3 = Ph$) are known to be sensitive to ring rupture under thermal and photochemical conditions [103, 104]. The species 2-formylindole, which would be analogous to the compound **9** shown in Scheme 3, is not reported. The nitrene intermediate that would potentially form in this particular case is an arylnitrene: these species rapidly form didehydroazepines through a very short lived benzazirine intermediate, and are known to polymerize [19].

Photoisomerization of quinoline *N*-oxide systems in methanol to generate quinolin-2 (1*H*)-one systems (**28/28′**) has been reported in the recent synthetic literature proceeding at moderate to high yields (75–89%) [105]. Generally, such isomerization systems are hypothesized to proceed via zwitterionic rather than biradical intermediates, since the formation of the quinolin-2(1H)-one is more readily explained by a 1,2-hydride shift in the zwitterion, rather than a 1,2-hydrogen atom shift in a 1,3-diyl biradical, for which there is more limited literature precedent. Similar reactions have been observed for benzimidazole *N*-oxides in alcohols [106], in addition to small amounts of products arising from ring scission and deoxygenation.

A study of the excited states of 6-hydroxyquinoline *N*-oxides indicates that these states have a high degree of acidity relative to the ground state, and thus these species may be characterized as photoacids. This reversible deactivation of the

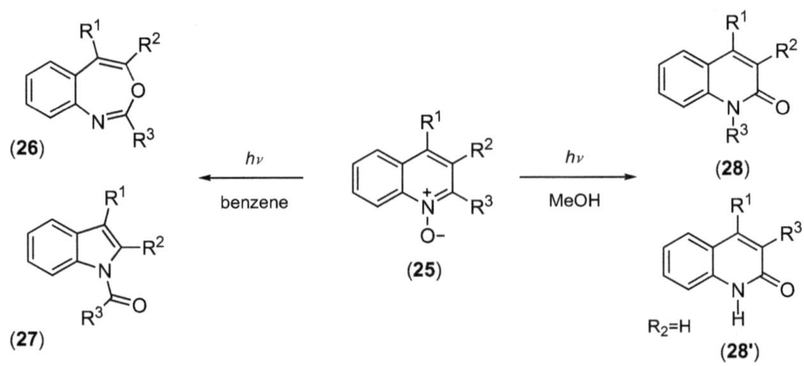

Scheme 15 Photochemistry of quinoline *N*-oxides

Table 2 Observed photochemistry of quinoline *N*-oxides (Scheme 15)

Substrate	Conditions	Yield (%) **28**/**28′**	**26**	**27** and hydrolysates	Deoxygenate
$R^1 = R^2 = R^3 = H$	Cyclohexane [108]	20	–[a]	58	
	Ether, acetone [50]	15	–	44	7
	Methanol [109]	60–70	–	3	5–6
$R^1 = R^2 = H, R^3 = CH_3$ [109, 110]	Methanol	16/22	–	10	4–5
$R^1 = R^2 = H, R^3 = CN$ [111]	Acetonitrile		81[b]		
$R^1 = R^2 = H, R^3 = CF_3$ [112]	Ether	–	70	–	3–5
	Methanol	–		70	3–5
$R^1 = R^2 = H, R^3 = OCH3$ [102]	Cyclohexane		–[c]	64	
	Water			75	
$R^1,R^3 = CH_3, R^2 = H$ [111]	Methanol	20/15	–	6–7	7–8
$R^1 = CH_3, R^2 = R^3 = H$ [111]	Methanol	70–75	–	5	1–2
$R^1 = Cl, R^2 = R^3 = H$ [111]	Methanol	75–80	–	–	2–3
$R^1 = OCH_3, R^2 = R^3 = H$	Cyclohexane [101]	10	–	45	–
	Methanol [111]	60–70	–	–	4–5
	Water [102]	95	–	–	–
$R^1 = CO_2H, R^2 = R^3 = H$ [113]	Methanol	79	6[b]		
$R^1 = CO_2H, R^2 = H, R^3 = CH_3$	Methanol [106]	20/31	37[b]		
$R^1 = H, R^2,R^3 = -(CH_2)_4$- [114]	Benzene	–		70[d]	5–10
	Methanol	34–48	–	–	5–10

[a]60% Yield of **26** if extraction is used rather than chromatography
[b]The oxazepine is believed to ring contract to yield the 3-formylindole
[c]90% yield of **26** if extraction is used
[d]R^2,R^3 form a spirocyclic ring at the 3-position. The formed cation is quenched by the alcohol

excited state competes with the more familiar deoxygenation reactions, and conversion to quinolin-2(1*H*)-ones [107].

Lei et al. report the synthesis of the antitumor agent 7-ethyl-10-hydroxycamptothecin **30** via photoisomerization of the *N*-oxide precursor **29** under acidic conditions [115]. In this case, group migration to effect ring expansion is prohibited by the geometric constraints of the planar ring. Similar outcomes (including regiochemical outcomes) were observed for 2-cyanoquinoline 1-oxide in acidic aqueous solutions [116]. In acidified alcohols, alkoxylation was observed.

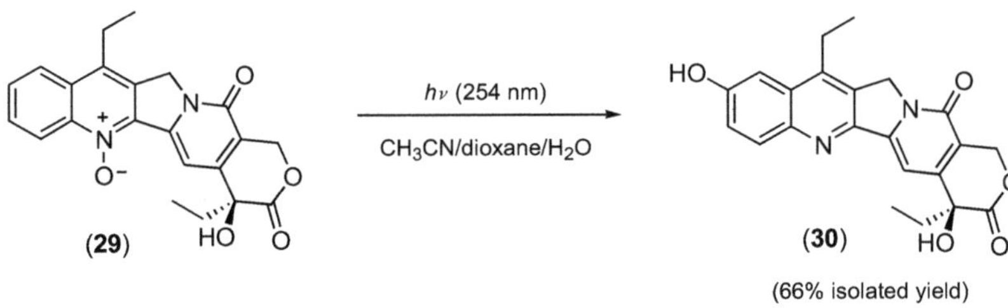

The compound 4-nitroquinoline 1-oxide (**31**) has received a large amount of attention due to its function as a model carcinogen. Nanosecond LFP of 4-nitroquinoline 1-oxide indicated the formation of a triplet species that absorbs at 410 and 560 nm [117, 118], and which underwent electron transfer and subsequent protonation with donors such as tryptophan, guanosine, and tyrosine to yield a radical **33** with a strong absorbance at 460 nm (Scheme 16). The efficiency of the electron transfer process for the triplet state is pH dependent – the rate of triplet quenching by tryptophan and tyrosine decreases with decreasing pH in aqueous solutions [119]. Choudhury and Basu generalized the process shown in Scheme 15 for **31** and indoles, and found that proton transfer could occur from the cation radical of the indole even if the indole was *N*-methylated [120]. HF and CASSCF calculations on the hydrogen-bonded complex of 4-nitroquinoline 1-oxide and tryptophan indicate that the excited singlet state of the complex is also a charge-transfer state that in principle may lead to the formation of **32** [121].

The intermediates **32** and **33** were observed by Shi and Platz [122] in a combined LFP/ns time-resolved infra-red (TRIR) study using DABCO as the donor species. The authors noted that under such conditions the formation of the deoxygenated 4-nitroquinoline is quenched, making the formation of 4-nitroquinoline from the expulsion of hydroxyl radical from **33** unlikely.

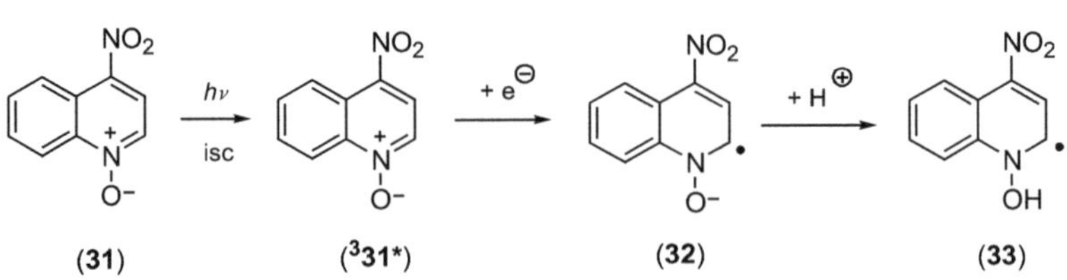

Scheme 16 Photochemistry of 4-nitroquinoline 1-oxide

The photochemistry of quinoxaline *N*-oxide **34** (and the bis-*N*-oxide, see below) has been the subject of a review [123]. These systems may be capable of generating isolable benz[d]-3,1,5-oxadiazepines (Scheme 17), as well as quinoxaline. The presence of methoxy groups at R^1 or R^2 leads to ring opening [124].

Phenanthridine *N*-oxides generate phenanthridones, deoxygenated phenanthridines, and the 1,3-oxazepine analogue (which may hydrolyze during chromatographic isolation, Scheme 18) [125]. In the specific case of the 6-cyano species (R^1 = CN), irradiation in acetone yields isolable amounts of 1,3-oxazepine (78%), phenanthridinone (analogous to **28**), and deoxygenated phenanthridine (3%) [126, 127]. Biradical intermediates may be trapped by alkenes [117]. In ethanol, *N*-ethoxyphenanthridone is the major product, and the presence of triplet sensitizers enhances the formation of the deoxygenated product. In an LFP study, the deoxygenation of the 6-cyano compound occurred via its T_1 state – the authors concluded that a triplet mechanism operates for deoxygenation of this species [128].

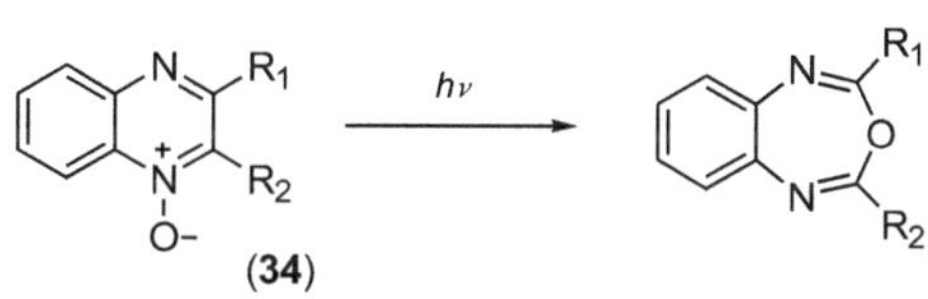

Scheme 17 Photochemistry of quinoxaline *N*-oxides

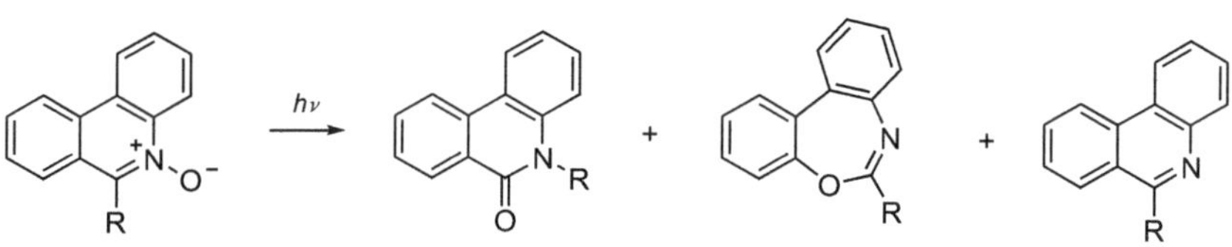

Scheme 18 Photochemistry of phenanthridine *N*-oxides

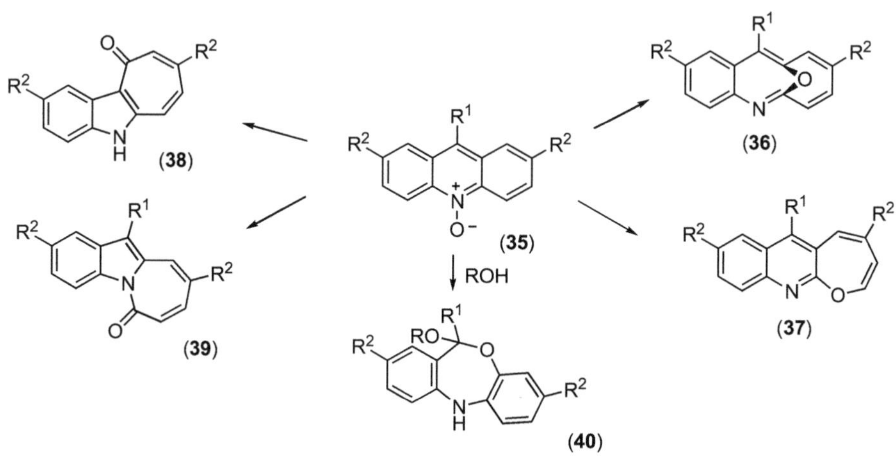

Scheme 19 Photochemistry of acridine *N*-oxides

The photochemistry of acridine *N*-oxides (Scheme 19) is believed to proceed via the rearrangement of the 1,3-oxazepine analog [129]. C–O bond scission could in principle yield a biradical or nitrile ylide intermediate. As with the pyridine and quinoline systems, the presence of a cyano group at the 10-position provides a certain degree of stabilization that allows small amounts of the 1,3-oxazepine **36** to be isolated, but the major components of the isolate are systems analogous to **28** (**39**) and products of ring expansion into the adjacent rings [130, 131]. Compound **40** is only observed in alcohol (R′OH, $R^1 = R^2 = H$) solutions [132].

Isoquinoline *N*-oxide species are similar in many respects to their quinoline analogues (Scheme 20), in that they commonly generate mixtures of oxazepine and isoquinolinones, with the former susceptible to hydrolysis [133].

LFP of isoquinoline (**41**) and benzocinnoline (**42**) *N*-oxides generate triplet species, which undergo an apparent hydrogen atom transfer reaction with hydroquinone with rate coefficients of $7 \times 10^6\ M^{-1}\ s^{-1}$ at 298 K, but react with well-known hydrogen atom donors such as $((CH_3)_3Si)_3SiH$ at much slower, experimentally unresolvable rates [122]. Interestingly, DFT calculations indicate that a hydrogen atom transfer reaction is thermochemically favorable for the hydroquinone system, whereas both electron transfer and proton transfer are not.

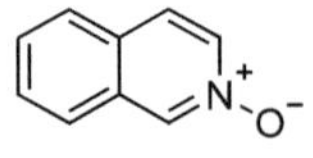

(**41**)

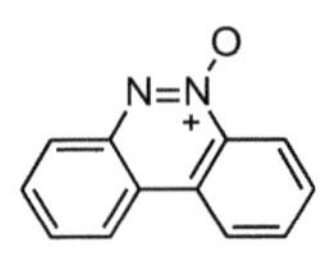

(**42**)

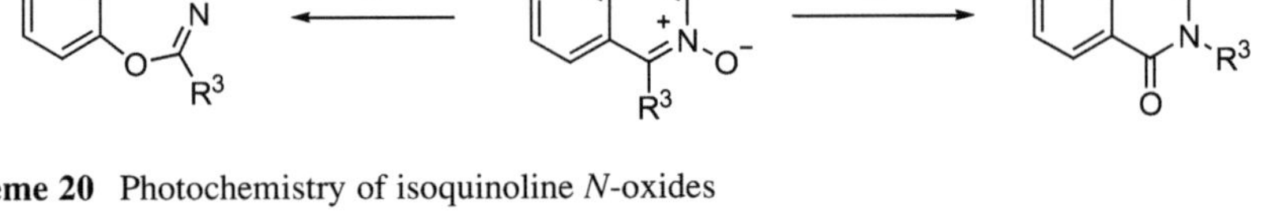

Scheme 20 Photochemistry of isoquinoline *N*-oxides

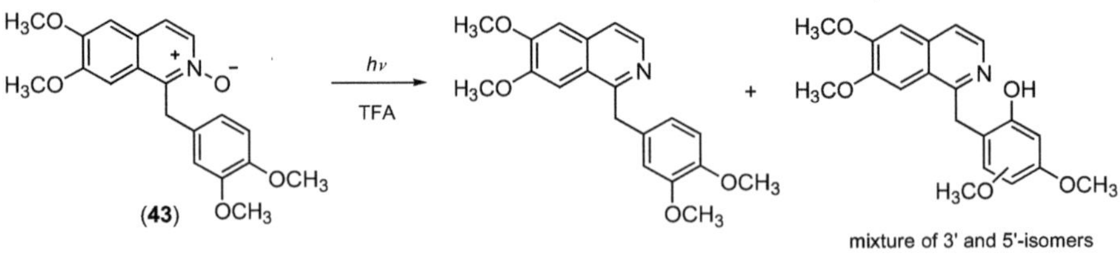

Scheme 21 Photochemistry of papaverine *N*-oxide in acidic conditions

The isoquinoline compound papaverine *N*-oxide (**43**), which may be described as a donor-spacer-acceptor system, will photorearrange under acidic conditions to generate hydroxylated derivatives of papaverine (Scheme 21) [134–136]. The authors postulated that hydroxylation would occur via a protonated singlet state, which would involve intramolecular charge transfer from the donor moiety to the acceptor (the isoquinoline *N*-oxide). The degree to which this occurs is dependent on the donor ability of the pendant ring, and the selectivity of reaction could be described by considering the spin densities of the radical cation corresponding to the donor moiety [137, 138]. Deoxygenation was postulated to occur via the protonated triplet state, which was confirmed by triplet quenching and triplet sensitization experiments performed by Collado et al. on a range of structurally similar systems [137].

The spirocyclic aziridine from 1,2-benzisoxazole, which could in principle be analogous to an intermediate formed from isoquinoline *N*-oxide photochemistry has been observed experimentally [43].

Pteridine *N*-oxides undergo facile deoxygenation reactions to yield pteridines [139]. The compound pyrimido[4,5g]pteridine 10-oxide (**44**) has been the subject of considerable study as a photochemical oxidant with reports of the photooxidation of styrene [140], tryptophan derivatives (at the β-carbon) [141], phenols (in addition to isoalloxazine *N*-oxide **45**, and pyridazine *N*-oxide) [142] and aromatic hydrocarbons [143]. In addition, photocyclizations of nucleosides [144] and chalcones [145] and dealkylation of *N,N*-disubstituted anilines [146] have been achieved with this compound. In all of these processes, the authors propose a mechanism whereby the excited state of **44** undergoes electron transfer to yield the radical anion of **44** and a radical cation species. Protonation of the radical anion in principle leads to a species that will fragment to form hydroxyl radical. In each case, radical cation chemistry yields the pteridine derivative as the major photoproduct.

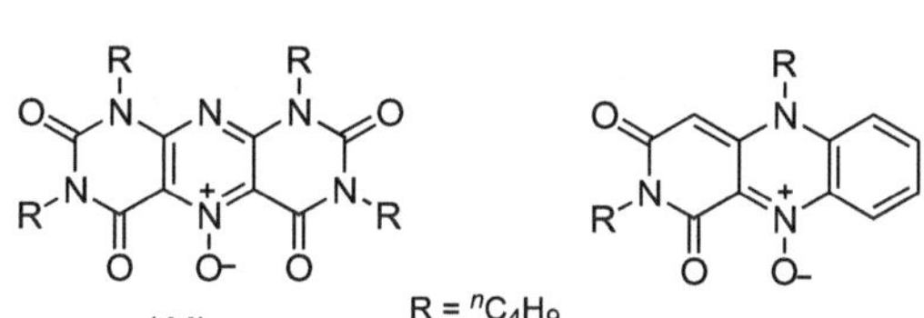

(**44**) (**45**)

This oxidant system has been coupled with an acridine based intercalator to generate a species capable of strand breakage of DNA. The quenching of strand breakage in the presence of DMSO lends support to the notion that hydroxyl radical is the oxidizing species [147].

2.3 Bis-N-*Oxide Systems*

To a significant extent, the photochemistry of bis-*N*-oxides is similar in many respects to those of their mono-*N*-oxide analogues. Pyrazine, quinoxaline, and phenazine *N,N*-dioxides are all capable of photochemically oxidizing amines to aminoxyl radicals [148, 149]. EPR evidence suggests the involvement of aminoxyl free radical intermediates, possibly arising from the triplet excited state [150, 151]. In aqueous solutions, pyrazine 1,4-dioxide will isomerize to yield either the 2,3- or 2,5-dihydroxypyrazines (tautomeric forms of the nominal pyrazinones) [152].

However, there are some reactions that are more specific in their outcomes. Pyridazine 1,2-dioxides were initially reported to undergo dual ring expansion reactions at low to moderate yields to generate 1,4,6,7-dioxadiazocins [153], which were subsequently identified and characterized as 3a,6a-dihydroisoxazolo [5,4d]isoxazoles (Scheme 22) [154]. The reaction is believed to proceed via N-N bond fragmentation, and the cyclization of the formed bis(iminoxyl) biradical.

The bicyclic bis-*N*-oxides have received a great deal of attention due to their possible clinical application. Quinoxaline bis-*N*-oxides, used as growth promoters in animal husbandry have been shown to undergo photodeoxygenation reactions [155], which can lead to modification of proteins [156]. In addition, these species have been shown to undergo conversion into quinoxalinone oxide, in a manner presumably analogous to the quinoline systems (Scheme 23). Pendant phenyl groups adjacent to *N*-oxide moieties may also be *ortho*-hydroxylated [157]. Aroyl substituted systems (R^1 = benzoyl, R^2 = phenyl) undergo additional photochemistry to yield benzimidazolones [158–160].

As with the simple pyridine 1-oxides, aryl substitution ($R^1 = R^2$ = phenyl, p-$MeOC_6H_4$) provides sufficient stability that 1,3,6-oxadiazepine 6-oxides may be isolated in small yields, but the primary photoproducts are 1,2-dibenzamides [161].

One of the most recently studied of the bis-*N*-oxides systems is the compound 3-aminobenzo1,2,4-triazine 1,4-dioxide (tirapazamine **46**), a compound implicated

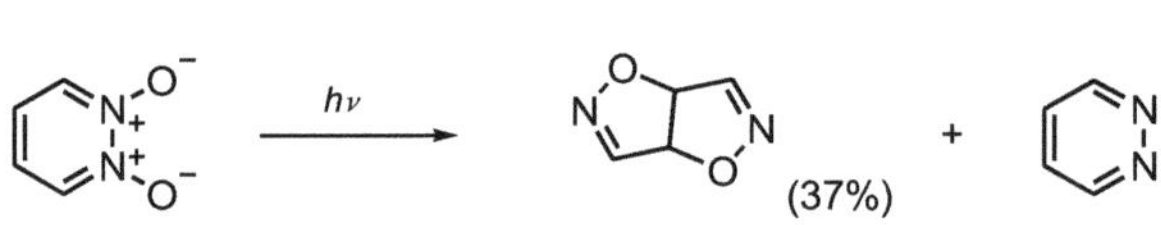

Scheme 22 Photochemistry of pyridazine 1,2-dioxide

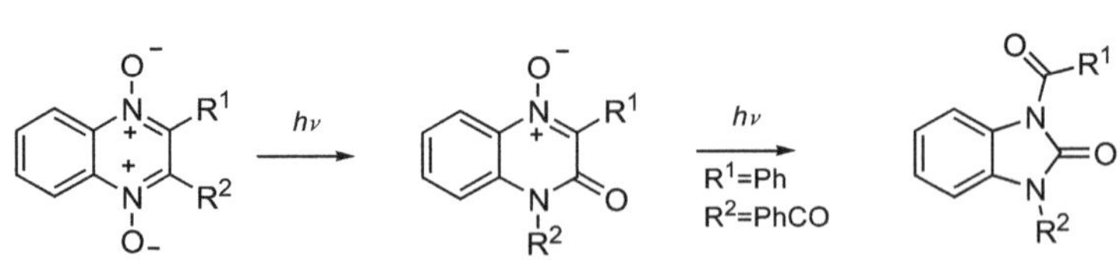

Scheme 23 Photochemistry of quinoxaline bis-*N*-oxide

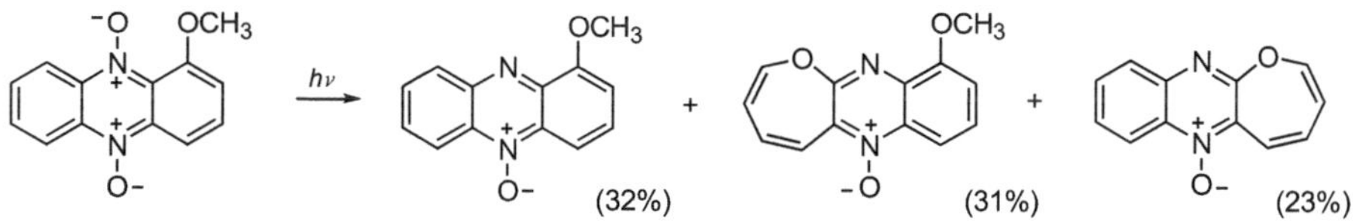

Scheme 24 Photochemistry of methoxyphenazine *N,N*-dioxide

as a possible treatment modality for hypoxic tumors [162, 163]. Fluorescence quenching studies indicate that the singlet excited state [164] of this compound, as well as the desoxy species **47** may undergo electron transfer with a range of potential donors at diffusion-limited rates to generate radical anions [165, 166]. These radical anions may promote photochemical DNA cleavage, although it was originally thought that this was achieved by protonation of the radical anion and subsequent expulsion of hydroxyl radical [167].

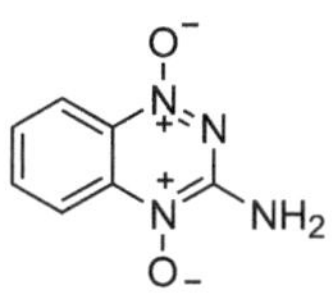
(46)
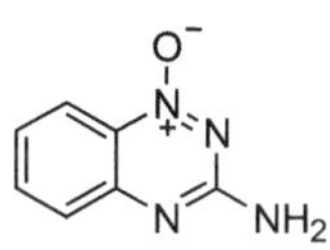
(47)
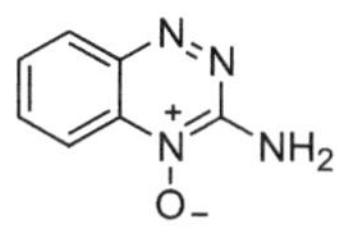
(48)
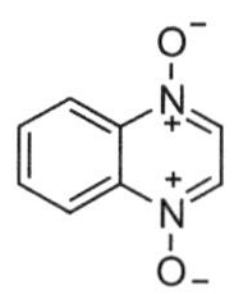
(49)
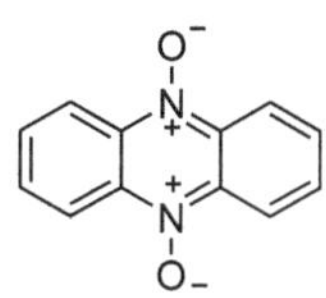
(50)

Shi et al. studied the singlet state lifetimes of the tirapazamine-derived species **46–48**, as well as the *bis-N*-oxides quinoxaline di-N-oxide (**49**) and phenazine di-N-oxide (**50**) [164]. These workers determined that the species **47** had an unusually long singlet lifetime (5.4 ns, compared to 110–130 ps for **46**). DFT calculations indicated that cyclizations to form oxaziridines were energetically favorable processes (i.e., $\Delta H_{R,cyc} < E(S_1)$) that shorten the singlet lifetimes, whereas cyclizations to generate the oxadiaziridines (the only cyclization process available to **47**) were not.

Phenazine *N,N*-dioxides have photochemistry similar to their acridine analogues: photolysis of methoxyphenazine *N,N*-dioxide yields a mixture of mono-deoxygenated product, as well as a pair of cyclized compounds structurally analogous to **35** [168] Scheme 24.

In organic solvents, the isolated product reported by Kawata et al. is assigned to the 1,3,5-oxadiazepine 3-oxide species that would arise from ring expansion in the central ring, analogous to the species observed in acridine *N*-oxides [169].

3 *N*-Alkoxyheteroarenium Salts

The photochemistry of the *N*-alkoxy salts is dominated by the fragmentation of the N–O bond to yield alkoxyl radical species [170]. Direct photolysis of *N*-ethoxyisoquinolinium ion generates ethoxyl radical and the isoquinoline radical cation [171], which may undergo electron transfer reactions with suitable donors, such as thymine dimers [172].

Studies by Collado and Perez reported that benzylated *N*-alkoxyisoquinolinium systems (acceptor-spacer-donor systems) when photolyzed would generate products consistent with alkoxyl radical generation and reaction with the rings [173]. The authors postulated two competing photoreactions, depending on the presence and the donor abilities of the pendant ring: in the absence of the pendant group, or where the group is a poor donor, substitution occurs on the isoquinoline ring following direct homolysis of the C–O. Where the group is a good donor, substitution occurs on the pendant ring following electron transfer (Scheme 25).

Direct photolysis of *N*-methoxyphenanthridinium cation is believed to generate phenanthridinium radical cation(λ_{max} = 670 nm) via N–O bond scission [174], which in turn is capable of electron transfer reactions [175]. This chemistry has been used to generate sulfide radical cations for LFP studies [176–179].

Kochi et al. demonstrated that fragmentation of these species may be facilitated via photoinduced charge transfer interactions between pyridinium salts and donor species such as arenes [180–182] to yield products consistent with alkoxyl radical attack on the arenes themselves. The general reaction is shown in Scheme 26.

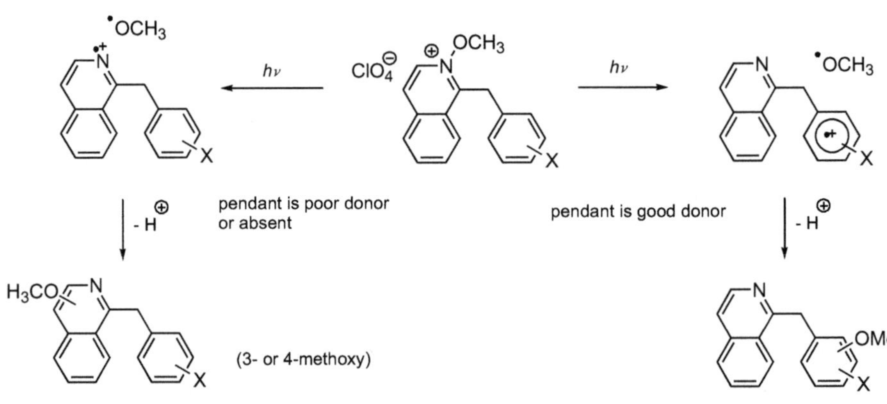

Scheme 25 Photochemical alkoxylation mediated by *N*-alkoxylisoquinolinium salts

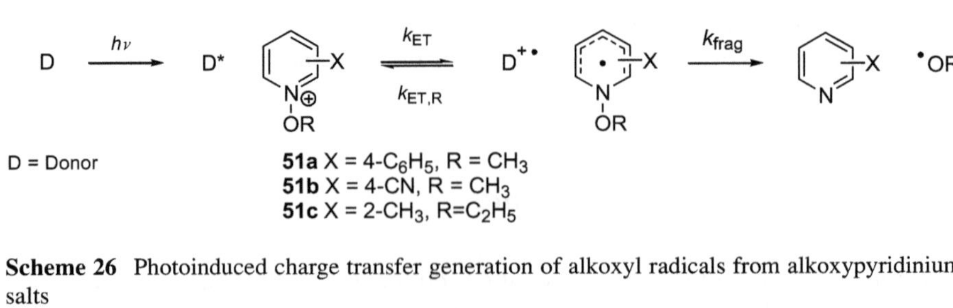

51a X = 4-C_6H_5, R = CH_3
51b X = 4-CN, R = CH_3
51c X = 2-CH_3, R=C_2H_5

Scheme 26 Photoinduced charge transfer generation of alkoxyl radicals from alkoxypyridinium salts

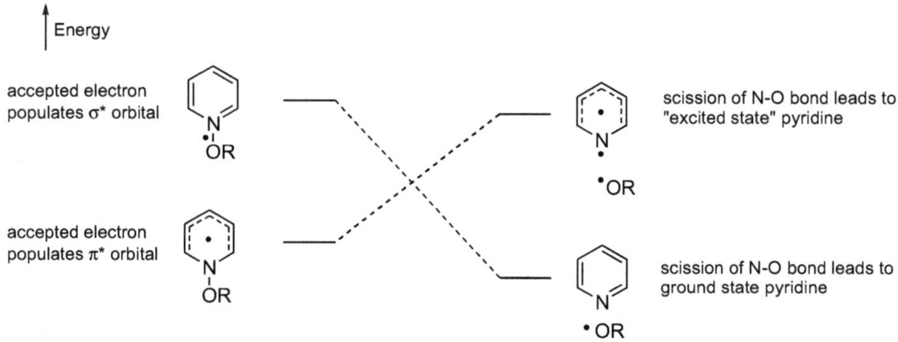

Scheme 27 Simplified state correlation diagram for fragmentation of *N*-alkoxypyridyl systems

Similar chemistry may be observed for the pyridinium salt **51b** with 1,2,4-trimethoxybenzene [183].

Gould et al. demonstrated that electron transfer could be achieved using a range of dye/*N*-methoxy-4-phenylpyridinium tetrafluoroborate (**51a**) pairs, where the dye absorbance maxima spanned the visible spectrum [184]. Electron transfer reactions are thermochemically favorable under circumstances where the reduction potential of the donor $E_{red,D}$ is comparable to or more negative than that of the pyridinium salt: $E_{red,py}$, ca. −1 V (SCE). Thus, dyes with $E_{red,D} < -0.7$ V(SCE) will generate radicals with **51a**, but the efficiency reaches a maximum with dyes where $E_{red,D} < -1.1$ V(SCE). The approach has been broadened by Podsiadly et al. to include photoreduction of **51a**.BF_4 and **51c**.PF_6 with a range of dyes absorbing in the 380–540 nm region of the spectrum [185–189].

Lorance et al., in a sequence of combined experimental and computational papers, investigated the importance of conical intersections on the fragmentation rate of the formed *N*-alkoxypyridyl radical. In the simplified correlation diagram shown in Scheme 27, a conical intersection exists between aπ*-type state and a σ*-type state, however a crossing from one state to the other is symmetry forbidden in the planar molecule [190]. In bent systems, where the nitrogen is allowed to pyramidalize, a greater degree of state mixing is accommodated, and the reaction becomes increasingly adiabatic in nature, with a lowering of the potential barrier with greater degrees of pyramidalization [191]. In systems where the pyridine rings are substituted with donating groups, the reaction becomes effectively barrierless [192].

Katbatc and Pczkowski developed a photoinitiation system whereby two radicals may be generated if triphenylbutylborate is present as a counterion to the pyridinium salt [193, 194]. The system may be further modified to include *N*-alkoxypyridinium moieties tethered to the dye framework [195], as well as 2,2′-bispyridinium systems **52a** and **52b** [196].

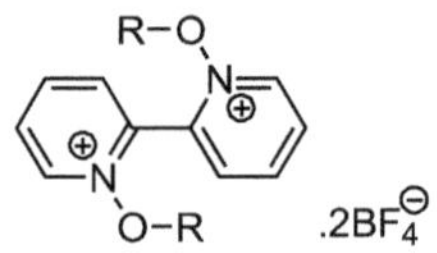

52a R = CH_3
52b R = C_2H_5

Significant chain amplification in photoinitiation systems may be achieved by the addition of reagents that themselves will act as donor systems. Thus, in dye/pyridinium (**51a**,**b**)/benzyl alcohol systems described by Shukla et al. [197, 198], the formed hydroxybenzyl radicals may act as donors, amplifying the observed quantum yield of fragmentation (Scheme 28). The protonated ketone is rapidly deprotonated by the formed pyridine. Similar enhancements may be observed in the presence of species such as lutidine, where it is believed that pyridinyl radicals act as the electron donating intermediates for *N*-methoxy and *N*-ethoxypyridinium systems [199]. This approach may be used synthetically to from benzo[b]phosphole oxides from phosphine oxides and alkynes, photoinitiated by a combination of *N*-ethoxypyridinium and Eosin Y – the formed

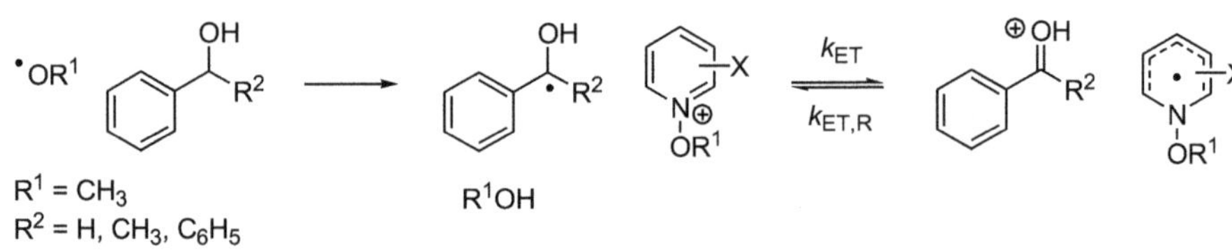

Scheme 28 Amplification of de-alkoxylation of *N*-alkoxypyridinium systems by benzyl alcohols

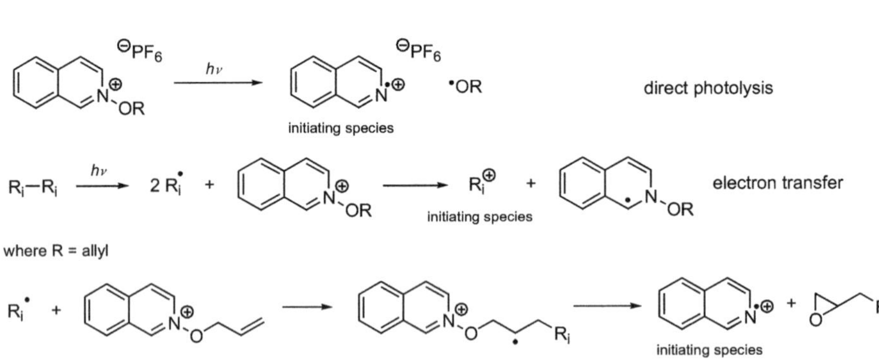

Scheme 29 Photoinitiation of cationic polymerization with *N*-alkoxylisoquinolinium salts

cyclohexadienyl radical acting as an electron-transfer agent in a manner analogous to that shown in Scheme 28 [200].

In addition to photoinduced charge transfer processes [201], alkoxyheteroarenium salts may also react with thermally or photochemically produced radical species to generate cations [202]. Thus, Yagci et al. reported that cationic polymerization may be initiated by the reaction of isoquinolinium systems with radicals to generate isoquinolinium radical cations (Scheme 29, where R_i refers to a free radical initiator fragment) [203]. Similar behavior was observed in the *N*-allyloxypyridinium analogues [204–208]. Even simple *N*-methoxy and *N*-ethoxypyridinium salts will behave similarly via direct photolysis or electron transfer with a range of potential free radical initiators [209–211].

As a third alternative means of photoinitiation, Arsu et al. demonstrated that alkoxypicolinium salts will generate picoline from photoinduced charge transfer reactions with anthracene. The generated picoline will polymerize cyanoacrylate monomers via so-called zwitterionic polymerization [212].

In addition to *N*-alkoxy salts, it may be shown that *N*-acyloxyheteroarenium species may also undergo N–O bond scission either by direct photolysis [213] or mediated by charge-transfer chemistry in the presence of suitable donors [180].

4 *N*-Hydroxypyridone and Its Derivatives

N-hydroxy-2-pyridone (**2**, R = H) may be considered a tautomeric form of 2-hydroxypyridine *N*-oxide. Low temperature glass studies indicate that the triplet state of *N*-hydroxy-2-pyridone exists in both tautomeric forms, indicated by dual phosphorescence behavior [214]. Room temperature LFP at 308 nm of this compound in organic and aqueous (pH ≤ 7) phases leads to the formation of hydroxyl and pyridoxyl radicals (Scheme 30) with quantum yields ranging between 0.25 and 0.6 [215]. Deprotonation at high pH quenches hydroxyl radical generation, and photoionization of the anion is inefficient.

Photolysis of **2** generates the hydroxyl radical, and is considered a specific source of this intermediate, which causes oxidative strand breaking in DNA [216]. This moiety (and its alkoxy derivative, see below) has been incorporated into chlorin p6 analogues (**53**) to generate a phototoxic system that generates hydroxyl or singlet oxygen intermediates capable of causing cell damage [217].

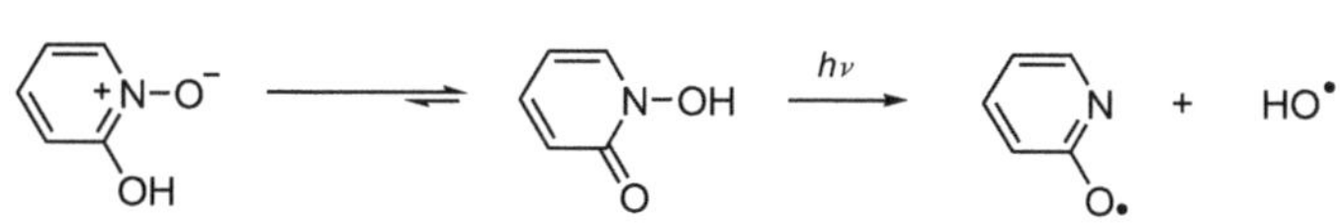

Scheme 30 Photochemistry of *N*-hydroxy-2-pyridone

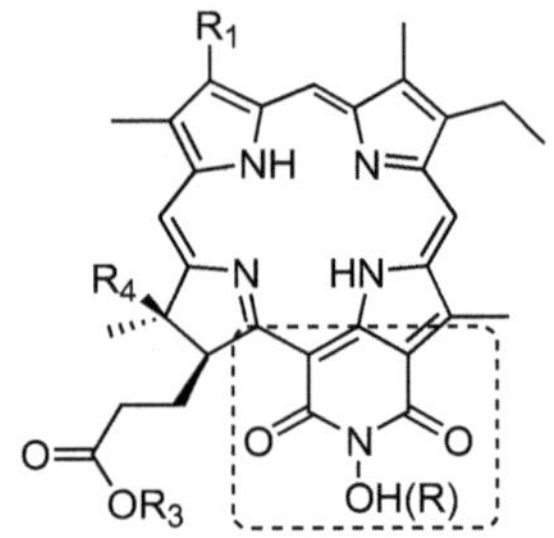

(53)

In principle, the photochemistry of the *N*-alkoxy and *N*-acyloxypyridones would be expected to mirror that of the *N*-hydroxy compound: N-O scission to yield a pyridoxyl radical and an alkoxyl radical (Scheme 31) [218]. Adam et al. utilized the isopropoxy derivative to generate isopropoxyl radicals in solution, which were capable of rearrangement in aqueous solution to yield the carbon-centered 2-hydroxy-2-propyl radical. Combination of the radicals with oxygen yielded peroxyl radicals, which were found to effective in cleaving DNA [219]. In a similar vein, the *N-tert*-butoxy derivative was utilized as a source of *tert*-butoxyl and methyl radicals to determine which of the species caused damage to supercoiled DNA by oxidation of deoxyguanosine residues [220, 221].

In the case of the 1-(9-anthrylmethoxy) pyridine, two modes of decomposition have been postulated (Scheme 32). In relatively non-polar environments, N-O

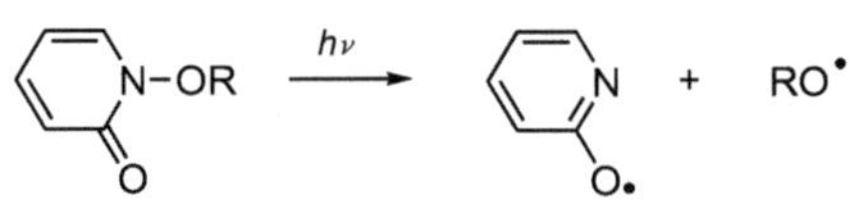

Scheme 31 Photochemistry of *N*-alkoxy-2-pyridones

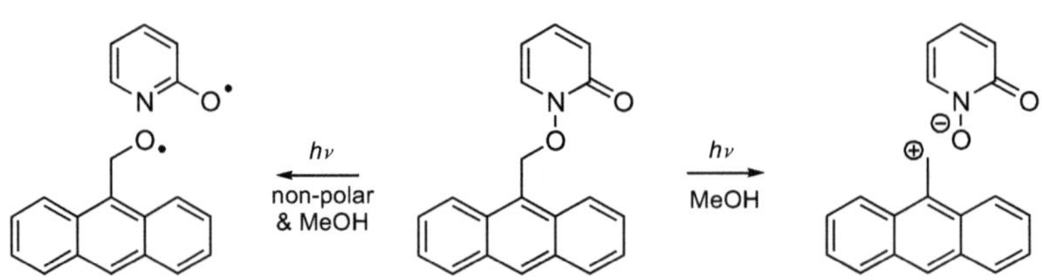

Scheme 32 Photochemistry of *N*-anthrylmethoxy-2-pyridones

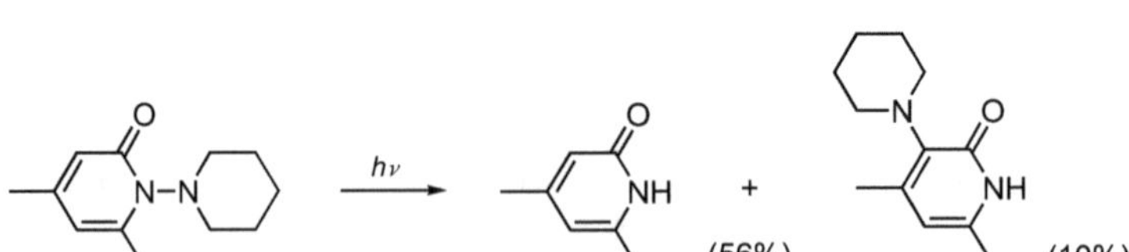

(54) (56%) (55) (10%)

Scheme 33 Photochemistry of *N*-amino-2-pyridones

scission dominates to yield an alkoxyl radical [222]. In methanol solvent, heterolytic C–O bond scission, hypothesized to proceed via an internal charge transfer exciplex, was observed to be competitive with N–O bond homolysis in 9-anthrylmethyl and pyrenyl derivatives [223].

In addition to the formation of alkoxyl radicals, aminyl radicals may also be obtained via the photolysis of *N*-amino species such as **54** (Scheme 33) [224]. The product **55** is considered to be formed by reaction of the aminyl radical with the pyridoxyl radical, or another pyridine. Similar products corresponding to alkoxyl radical attack or carbon centered radicals arising from fragmentation of alkoxyl radicals have been reported [225, 226].

Acyloxy radicals may be generated from the appropriate hydroxamic ester (**56**) [227]. An alternative approach has been used to generate acyloxy radicals from pyrazinones (**57**) [228] or purines (**58**) [229, 230].

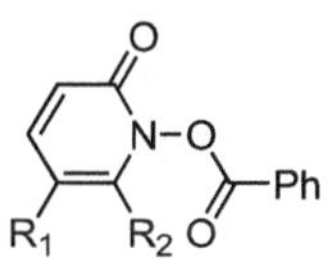

(**56**)

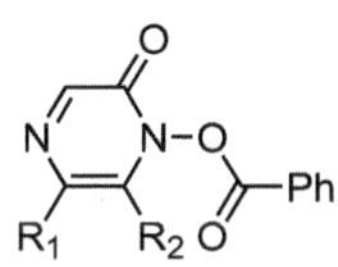

(**57**)

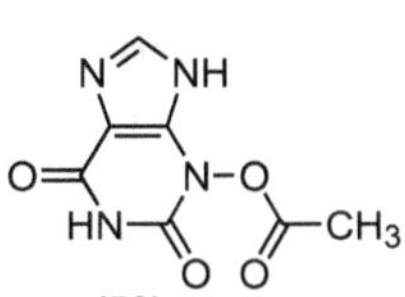

(**58**)

Although the 2-pyridone derivatives exhibit relatively clean photochemistry, the bulk of free radical chemistry arising from these types of derivatives has centered around the pyridine-2-thione (pyrithione) derivatives, possibly due to their greater thermal lability, but also to their more specific absorbance window.

5 *N*-Hydroxypyrithione and Its Derivatives

5.1 N-*Hydroxy-2-Pyrithione and* N-*Alkoxy-2-Pyrithiones*

As with its pyridone analogue, *N*-hydroxy-2-pyrithione (NPT, **3**, R = H) may be thought of as a tautomeric form of 2-mercaptopyridine *N*-oxide. The photochemistry of this species is somewhat more complex than its pyridone analogue: Aveline et al. [231] reported that laser flash photolysis of this species does generate the expected hydroxyl and pyrithiyl radical **59** in organic solvents (Scheme 34), by analogy with Scheme 30 ($\Phi = 0.3$–0.45), along with a small amount ($\Phi = 0.03$–0.05) of triplet state that does not undergo electron transfer under the experimental conditions. The generated pyrithiyl radical is a relatively stable species, and will ultimately dimerize. Botchway et al. have demonstrated that the photodissociation may also be achieved by fast two-photon processes, when solutions of NPT are irradiated at 750 nm [232]. The relatively clean photochemistry of this species in organic solvents allows it to be used as a photochemical source of hydroxyl radicals for LFP competitive kinetic studies, using *trans*-stilbene as a hydroxyl radical reporter species [233–236].

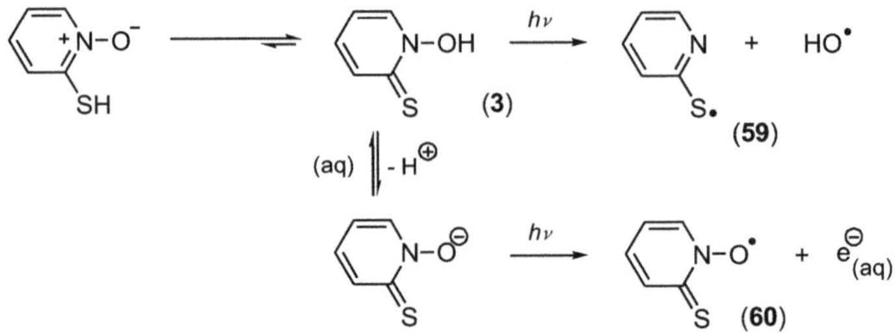

Scheme 34 Photochemistry of *N*-hydroxy-2-pyrithione (NPT)

In aqueous solution, Aveline et al. reported that *N*-hydroxypyrithione was present in its anionic form at pH 7, and that this species was also susceptible to photoionization, leading to the formation of solvated electron (Scheme 34), in competition with hydroxyl radical generation [237]. Photolysis at pH 2 did not generate the solvated electron, but did yield the triplet state observed in organic media. Thus, this compound exhibits the potential to generate more than one type of reactive intermediate in aqueous solutions [238, 239]. A recent high-level configuration interaction (SAC-CI) computational study of the photoionization of NPT in solutions simulated by a polarizable continuum model (PCM) indicates that the thione form of NPT is approximately 8–10 kcal/mol more stable than the thiol form, depending on the solvent field [240]. The same study confirmed that the photoionization reaction is most likely associated with charge transfer between the anion and water.

The kinetics of both the pyrithiyl radical and its *N*-oxyl analogue (**60**, generated by photoionization, $\lambda_{max} = 365$ nm) were studied by the laser flash photolysis of their respective disulfides [241]. The authors found that in general **60** behaves as a thiyl radical – it dimerizes at diffusion controlled rates, and undergoes addition to alkenes, albeit at a rate significantly slower than the pyrithiyl radical observed from dissociation of NPT.

Much of the interest in NPT lies in its ability to generate hydroxyl radicals photochemically under biological conditions, to study their subsequent damage of nucleic acids [242]. This facet of reactivity has engendered a number of studies: Chaulk et al. studied RNA cleavage by photolysis of NPT as a potential hydroxyl radical "footprinting" methodology [243]. Douki et al. used NPT photolysis to study hydroxyl radical induced oxidation of deoxyguanosine [244], and Telo utilized NPT as a hydroxyl radical source for EPR studies of radical anions derived from reaction of hydroxyl with uric acids [245]. Vrantza et al. synthesized deoxyguanosine derivatives tethered to NPT groups: photolysis of these species led to the formation *N*-deoxygenated species, the loss of the pyrithione group, and glycosidic bond cleavage [246].

The zinc complex of NPT (**61**) is used as a marine anti-fouling agent, and as a component of various personal care products, and so its fate in natural waters is of interest. Sakkas et al. demonstrated that the primary products of photolysis of **61** in water under simulated solar irradiation were disulfide products with varying degrees of deoxygenation at nitrogen. Photolysis is accelerated by the presence of dissolved organic matter and nitrate [247].

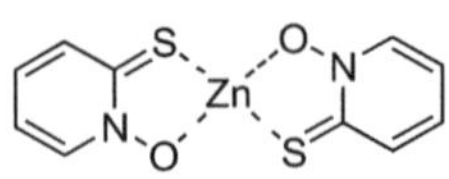

(**61**)

Aveline and Redmond also considered the photochemistry of the isomeric *N*-hydroxy-4-pyrithione [248]. The photochemistry of this compound is complicated somewhat by the fact that it exists in an equilibrium of tautomers (**62/62′**), the position of which is strongly dependent on solvent (the form **62** favored in less polar solvents, Scheme 35). Unlike the 2-isomer, photolysis of *N*-hydroxypyrithione **62** does not yield hydroxyl radical, and the photochemistry is dominated by triplet chemistry (including quenching by starting material, not shown in Scheme 35). Photolysis of the *N*-oxide form leads to photodissociation of the S–H bond.

The *N*-alkoxyl-2-pyrithiones, by analogy with their hydroxyl analogues, also undergo photolysis with homolytic N–O bond cleavage to generate alkoxyl radicals, which are capable of guanine oxidation and ultimately DNA strand cleavage [249] and cell damage [250].

Arnone et al. have performed a number of theoretical studies in order to elucidate the mechanism of the photoreaction [251], and to design functionalized derivatives [252] or alternative structures of photochemical precursors [253] that avoid undesirable photolysis in ambient visible light. High-level CASPT2 calculations on *N*-methoxy-2-pyrithione suggest that dissociation occurs following a $\pi \rightarrow \pi^*$ transition to the S_2 state (the S_1 state arising from an $n \rightarrow \pi^*$ transition). The authors also indicate that in the case of the *N*-hydroxy analogue, dissociation occurs from the $\pi \rightarrow \pi^*$ S_1 state, due to

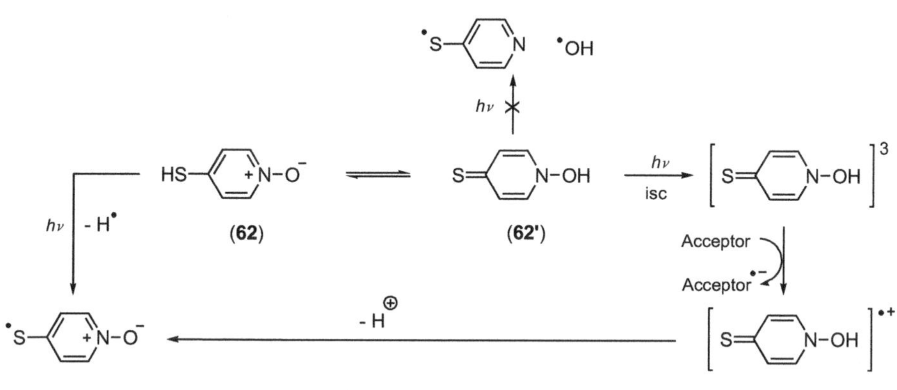

Scheme 35 Photochemistry of *N*-hydroxy-4-pyrithione

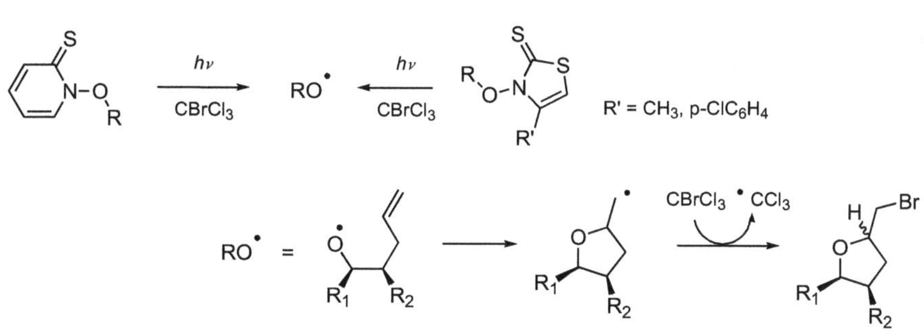

Scheme 36 Synthetic photochemistry of *N*-alkoxy-2-pyrithiones and *N*-alkoxythiazole-2(3*H*)-thiones

the stabilization of sulfur non-bonding orbitals by internal hydrogen bonding interactions with O–H [251].

The synthetic utility of these species has been demonstrated by Hartung et al., who photolyzed pent-4-enyloxy derivatives of NPT and structurally similar thiazole-2(3*H*)-thiones in $BrCCl_3$ to yield α-bromomethyltetrahydrofurans (Scheme 36) [254–257]. Hartung et al. demonstrated with chiral alcohol derivatives that the alkoxypyridinethiones could be generated with inversion of stereochemistry at the alcoholic carbon [258].

5.2 Barton Esters

The use of pyridine-2-thioneoxycarbonyl (PTOC) esters as carbon-centered radical precursors was developed so extensively by the group of Barton that these compounds are commonly referred to as Barton esters. These species are readily prepared from acyl halides [259], and thermally decompose to yield alkyl radicals. The photochemistry of these species closely mimics the thermal chemistry: photoexcitation leads to N–O bond scission with high quantum yield ($\Phi \sim 0.5$), and the (typically) rapid decarboxylation of the formed acyloxy radicals leads to the formation of alkyl radicals (Scheme 37). Both the pyrithiyl and carbon-centered radicals may attack the sulfur of the PTOC ester [260], the latter reversibly [261], leading to the generation of an additional equivalent of acyloxy radical.

Barton's group used these compounds to generate alkyl radicals (including radicals derived from amino acids) [262] capable of addition to a range of alkenes in both intra- [263] and inter-molecular modes [264–267]. Similar reactions may be observed for azo-compounds to generate hydrazine derivatives [268]. Alkylation of heteroarenes may be achieved via alkyl radical attack on their heteroarenium salts [269]. Alkyl radicals generated from Barton esters will attack xanthates, leading to an overall deoxygenation of alcohols (the Barton McCombie reaction) – the presence of a sulfone group on an adjacent carbon leads to formation of alkenes [270]. In a similar vein, carboxylic acids may be two-carbon homologated to α-ketocarboxylic acids via Barton esters [271]. Single carbon homologation is achieved by radical attack on isocyanides [272].

Certain acyloxy radicals will not decarboxylate rapidly, and will exhibit chemistry similar to alkyl radicals to yield sulfenyl esters [273]. Similar species were observed in matrix isolation studies of NPT, where photogenerated hydroxyl radical reacted with pyrithiyl to generate 2-hydroxysulfanylpyridine [274].

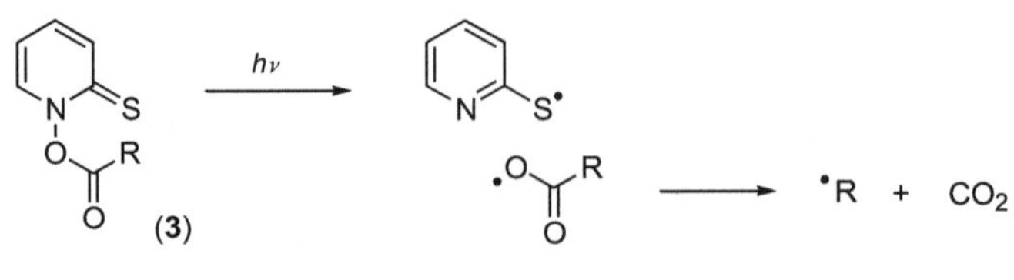

Scheme 37 Radical photochemistry of Barton esters

The photochemistry of Barton esters proceeds cleanly, with useful quantum yields [260, 275], making these esters useful free radical precursors for LFP studies (particularly as their absorption maxima are near 355 nm, making them accessible to photoexcitation by Nd-YAG lasers[276], although it can be shown that these species may also decompose via two-photon processes) [277]. In particular, Newcomb's group has used a number of Barton ester precursors for alkyl radicals as part of their LFP investigations of fast radical rearrangements [278–281], so-called radical clock reactions. Barton esters have been used as precursors of enol ether radical cations via fragmentation of β-phosphatoxy radicals (Scheme 38) [282], and have been used to establish radical cation "clock" cyclization and fragmentation reactions [283, 284]. Similar reactions occur for other good leaving groups, such as mesylates [285].

These same methods may also be used in product distribution studies: Boyles and Carpenter used the Barton ester to compare and contrast the chemistry observed for a free radical with that from a structurally similar singlet biradical [286]. Recently, Kawamoto et al. used Barton ester photochemistry to estimate the rate of hydrogen atom abstraction by alkyl radicals from cyanoborohydrides [287].

Barton esters may also be used to generate alkyl radicals from tellurides photochemically, where an alkyl radical generated by Barton ester undergoes a group transfer reaction with an alkyl aryl telluride (this reaction does not occur for sulfides and selenides) [288]. Attack of alkyl radicals formed from Barton esters on sulfur leads to the formation of thiols [289]. The photolysis of Barton esters in halogenated solvents (CCl_4, $CBrCl_3$ or CF_3CH_2I) will yield halides, in high yields in the case of bromination [290, 291].

Alkyl radicals photogenerated from Barton esters may be shown to cleave duplex [292], and supercoiled DNA [293], and Castagnino et al. have studied the photorelease of alkyl radicals from non-polar Barton esters incorporated in cyclodextrin hosts as a potential therapeutic vehicle [294]. Aveline and Redmond utilized Barton esters (**3**, R = benzyl, dimethylphenyl, *tert*-butyl) to assess their phototoxicity as carbon centered radical precursors, relative to NPT (see above) and a benzoyl derivative (**3**, R = Ph) as oxygen centered radical precursors, the latter due to the slow decarboxylation of the benzyloxyl radical [295]. The photochemistry of both alkyl (adamantyl) and aryl (nicotinyl) Barton esters was used to assess the radical scavenging properties of chitosan [296].

The photochemistry of Barton esters has been used synthetically to generate quinone sesquiterpene natural products, where the critical step is the addition of a photochemically generated sesquiterpene alkyl radical to a suitably substituted quinone [297]. In addition, Barton ester photolysis may be utilized to generate heavily halogenated ring systems by reacting but-3-enylradicals with CF_2CCl_2 [298], and fused oxirane ring systems may be obtained by photochemical

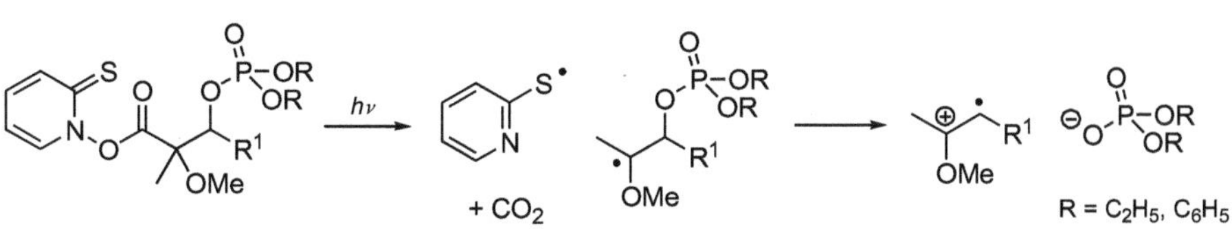

Scheme 38 Formation of radical cations from Barton esters

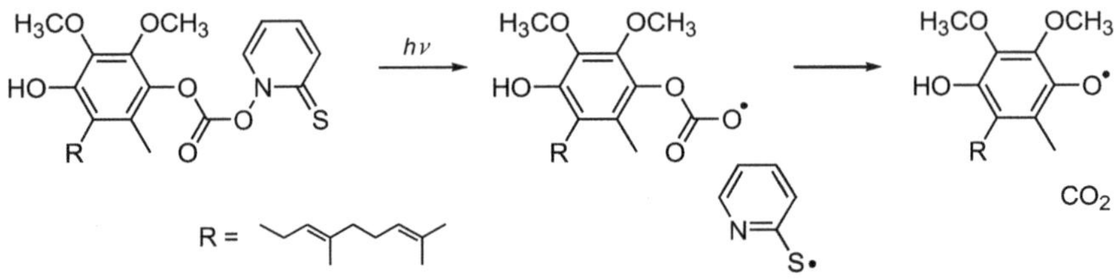

Scheme 39 Formation of oxygen-centered radicals from Barton carbonates

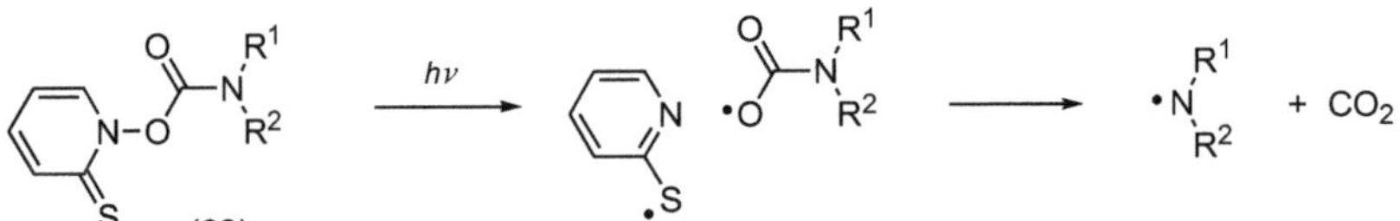

(63)

Scheme 40 Formation of nitrogen-centered radicals from Barton carbamates

generation of oxiranyl radicals [299]. The generation of enolate radical cations from photolysis of β-phosphatoxy Barton esters (Scheme 38), when trapped with allyl alcohol, provides a synthetic pathway to spirocyclic ethers via radical cyclization [300]. Incorporation of Barton ester groups on styrene leads to the formation of polymeric materials that can be used photochemically to generate the so-called hyperbranched copolymeric materials [301].

Species that might be thought of as Barton carbonates will yield carbonatyl radicals on photolysis. Decarboxylation of the carbonatyl radical is slower than that of the acyloxy radical, so much that these species may be trapped by internal cyclization onto alkenyl side chains to generate cyclic carbonates [302]. Fragmentation to generate phenoxyl radical is more facile, and Barton carbonates have been used as photochemical precursors of phenoxyl radicals [303], and semiquinones (including semiubiquinones, Scheme 39) [304]. These species disproportionate in solution to yield quinones and hydroquinones, although the authors indicate that under certain circumstances, a formal disproportionation may occur between semiquinone and pyrithiyl radical.

Barton carbamates (**63**) have been used to generate *N*-centered radicals (Scheme 40) [305]. Esker and Newcomb have photochemically generated amidyl radicals from carbamates (R^1 = alkyl, R^2 = acyl); where R^1 contains a double bond, cyclization is possible [306]. Burdi et al. have generated tryptophanyl radicals from the carbamate derivative **64** by LFP [307].

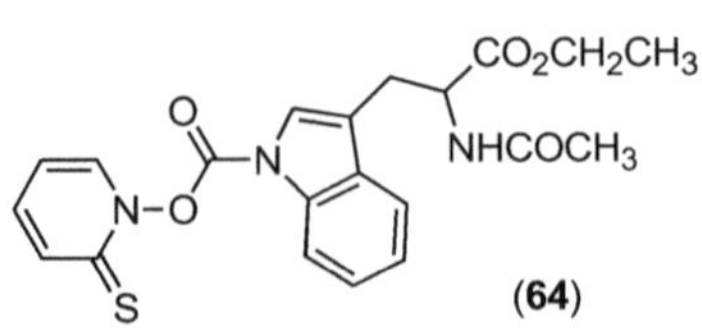

(64)

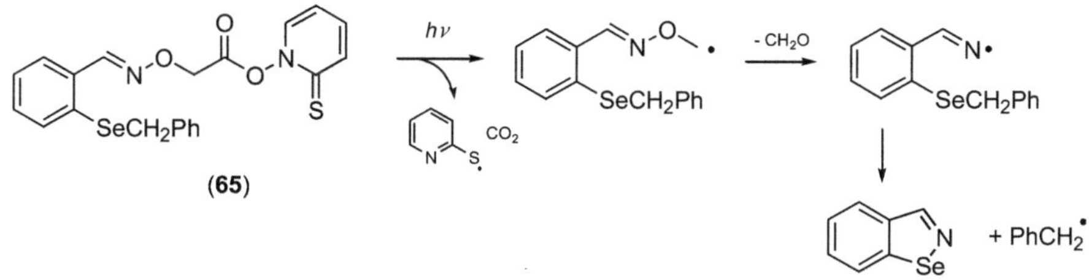

Scheme 41 Formation of benzoselenazole via imidyl radical intermediates

In an analog of earlier work involving the synthesis of selenoethers [308], Fong and Schiesser have generated an iminyl radical indirectly from photolysis of Barton ester **65**, followed by decarboxylation and loss of formaldehyde to generate 1,2-benzoselenazole (Scheme 41) [309].

In addition to pyridine-2-thiones, free radicals have also been photochemically generated from 1-acyloxy-2(1*H*)pyrimidine-2-thiones (**66**) [310].

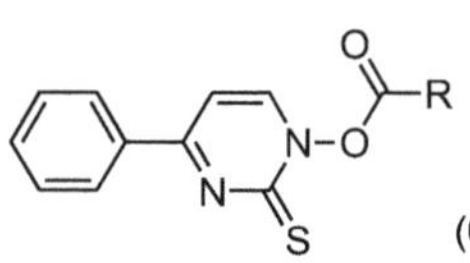

(66)

6 Concluding Remarks

As has been noted before, the photochemistry of heteroarene *N*-oxides is rich and varied, and there remain some mechanistic questions to be resolved. Some mechanistic insight may be gained from continuing mechanistic investigation of nitrene and azirine species. One of the issues yet to be resolved arises from the differences observed in photochemistry from aprotic to protic solvents, with particular emphasis on examples such as the fragmentation of pyridine *N*-oxides in basic aqueous solutions, and the changes in photochemistry observed for quinoline *N*-oxides observed in Scheme 15. Advances in computational methodologies, including the incorporation of solvation effects, may be able to provide some clarity as to the nature of the lowest singlet excited state for the *N*-oxides and possible intermediates such as **6** in non-polar, polar, and protic environments (with the additional caveat that the N-oxide group is capable of hydrogen bonding, and specific solvent-solute interactions may need to be considered). The oxidative capabilities of these species appear to arise from mechanisms arising from singlet fragmentation to yield $O(^3P)$, or to mechanisms involving electron transfer reactions.

On the experimental side, spectroscopic techniques such as laser flash photolysis and time resolved infra-red spectroscopy provide the ability to resolve the primary photochemistry of these species. Traditional product distribution studies have always been confounded by problems of secondary photochemistry, as well as instability of products upon workup.

The photochemistry of *N*-oxides and their derivatives plays a significant role in synthesis, polymer science, as well as medicinal and biological chemistry. Continuing development of compounds with these functional groups drives us toward understanding their photochemistry – a clearer picture has arisen over time, but there remain worthwhile problems to solve both experimentally and computationally.

References

1. Spence GG, Taylor EC, Buchardt O (1970) Chem Rev 70:231
2. Bellamy F, Streith J (1976) Heterocycles 4:1391
3. LaBlache-Combier A (1976) In: Buchardt O (ed) Photochemistry of heterocyclic compounds. Wiley-Interscience, New York, NY, pp 223–233
4. Albini A, Alpegiani M (1984) Chem Rev 84:43
5. Albini A, Fasani E, Amer AM (1995) In: CRC handbook of organic photochemistry and photobiology. CRC Press, Boca Raton FL, pp 879–891
6. Albini A, Fagnoni M (2004) CRC handbook of organic photochemistry and photobiology, 2nd edn. CRC Press, Boca Raton FL, pp 1–21
7. Dunkin IR (1998) Matrix isolation techniques – a practical approach. Oxford Univ. Press, New York, NY
8. Bally T (2004) In: Moss RA, Platz MS, Jones M (eds) Reactive intermediate chemistry. Wiley Interscience, Hoboken, NJ, pp 797–846
9. Scaiano JC (2004) In: Moss RA, Platz MS, Jones M (eds) Reactive intermediate chemistry. Wiley Interscience, Hoboken , NJ, pp 847–872
10. Bist HD, Parihar JS, Brand JCD (1976) J Mol Spectrosc 59:435
11. Bist HD, Parihar JS, Brand JCD (1977) J Mol Spectrosc 64:211
12. Peacock RD, Samori (1975) J Chem Soc Perkin Trans 2:1909
13. Scholz M (1981) J Prakt Chem 323:571
14. Kuzuya M, Noguchi A, Maki Y, Sako M (1991) Chem Pharm Bull 39:3110
15. Nakagawa Y, Suzuka I, Ito M (1993) Chem Phys Lett 208:453
16. Puszko A (1998) Khim Geterosik Soedin Engl Transl 34:1148
17. Palmer MH, Hoffmann SV, Jones NC, Smith ER, Lichtenberger DL (2013) J Chem Phys 138:214317
18. Parasuk V, Cramer CJ (1996) Chem Phys Lett 260:7
19. Platz MS (2004) In: Moss RA, Platz MS, Jones M Jr (eds) Reactive intermediate chemistry. Wiley-Interscience, Hoboken NJ, pp 501–560
20. Burdzinski GT, Middleton C, Gustafson TL, Platz MS (2006) J Am Chem Soc 128:14804
21. Hu B, DiMagno SG (2015) Org Biomol Chem 13:3844
22. Cao J (2015) J Chem Phys 142:244302
23. Galenko EE, Khlebnikov AF, Novikov MS (2016) Chem Heterocycl Comp 52:637
24. Pusch S, Schollmeyer D, Opatz T (2016) Org Lett 18:3043
25. Nunes CM, Reva I, Fausto R (2013) J Org Chem 78:10657
26. Sarkar SK, Osisioma O, Karney WL, Abe M, Gudmundsdottir AD (2016) J Am Chem Soc 138:14905

27. Rajam S, Jadhav AV, Li Q, Sarkar SK, Singh PND, Rohr A, Pace TCS, Li R, Krause JA, Bohne C, Ault BS, Gudmundsdottir AD (2014) J Org Chem 79:9325
28. Germeraad P, Moore HW (1974) J Org Chem 39:774
29. Hassner A (1984) In: Scriven EFV (ed) Azides and nitrenes – reactivity and utility. Academic Press Inc., Orlando, FL, pp 35–94
30. Germeraad P, Weyler W, Moore HW (1974) J Org Chem 39:781
31. Bellamy F, Martz P, Streith J (1975) Heterocycles 3:395
32. Bellamy F, Streith J (1979) J Chem Res 18
33. Chen F, Shen T, Cui Y, Jiao N (2012) Org Lett 14:4926
34. Alkaitis A, Calvin M (1968) J Chem Soc Chem Commun 292
35. Bellamy F, Streith J, Fritz H (1979) Nouveau J Chim 3:115
36. Buchardt O, Pedersen C, Harrit N (1972) J Org Chem 37:3592
37. Ishikawa M, Kaneko C, Yokoe I, Yamada S (1969) Tetrahedron 25:295
38. Cardillo P, Gigante L, Lunghi A, Zanirato P (2010) J Therm Anal Calorim 100:191
39. Xue J, Luk HL, Eswaran SV, Hadad CM, Platz MS (2012) J Phys Chem A 116:5325
40. Söderburg BCG (2000) Curr Org Chem 4:727
41. White RC, Wedyck EJ, Lynch KO (1993) J Photochem Photobiol A Chem 71:127
42. Nunes CM, Reva I, Pinhoe Melo TMVD, Fausto R (2012) J Org Chem 77:8723
43. Nunes C, Pinto SMV, Reva I, Fausto R (2016) Eur J Org Chem 4152
44. Liu Y, Guan P, Wang Y, Liu L, Cao J (2015) J Phys Chem A 119:67
45. Rajam S, Murthy RS, Jadhav AV, Li Q, Keller C, Carra C, Pace TCS, Bohne C, Ault BS, Gudmundsdottir AD (2011) J Org Chem 76:9934
46. Zhang X, Sarkar SK, Weragoda GK, Rajam S, Ault BS, Gudmundsdottir AD (2014) J Org Chem 79:653
47. Albrecht E, Mattay J, Steenken S (1997) J Am Chem Soc 119:11605
48. Naito I, Ishida A, Yamamoto K, Takamuku S, Isomura K, Taniguchi H (1989) Chem Lett 1615
49. Barcus RL, Wright BB, Platz MS, Scaiano JC (1983) Tetrahedron Lett 24:3955
50. Naito I, Ishida A, Takamuku S, Isomura K, Taniguchi H (1992) J Chem Soc Perkin Trans 2:1985
51. Mueller JO, Schmidt FG, Blinco JP, Barner-Kowollik C (2015) Angew Chem Int Ed 54:10284
52. Naito I, Morihara H, Ishida A, Takamuku S, Isomura K, Taniguchi H (1991) Bull Chem Soc Jpn 64:2757
53. Cludius-Brandt S, Kupracz L, Kirschning A, Beilstein J (2013) Org Chem 9:1745
54. Xuan J, Xia X-D, Zeng T-T, Feng Z-J, Chen J-R, Lu L-Q, Xiao W-J (2014) Angew Chem Intl Edn 53:5653
55. Zeng T-T, Xuan J, Ding W, Wang K, Lu L-Q, Xiao W-J (2015) Org Lett 17:4070
56. Becher J, Finsen L, Winckelmann I, Kognaty RR, Buchardt O (1981) Tetrahedron 37:789
57. Nakagawa M, Kaneko T, Yamaguchi H (1972) J Chem Soc Chem Commun 603
58. Finsen L, Becher J, Buchardt O, Koganty RR (1980) Acta Chem Scand B 34:513
59. Albini A, Fasani E, Lohse C (1988) Heterocycles 27:113
60. Lohse C, Hagedorn L, Albini A, Fasani E (1988) Tetrahedron 44:2591
61. Buchardt O, Christensen JJ, Lohse C, Turner JJ, Dunkin IR (1977) J Chem Soc Chem Commun 837
62. Buchardt O, Christensen JJ, Nielsen PE, Koganty RR, Finsen L, Lohse C, Becher J (1980) Acta Chem Scand B 34:31
63. Hassner A, Wiegand NH, Gottlieb HE (1986) J Org Chem 51:3176
64. Moore HW, Wilbur DS (1980) J Org Chem 45:4483
65. Moore HW, Hernandez L, Kunert DM, Mercer F, Sing A (1981) J Am Chem Soc 103:1769
66. Hegarty AF, Tuohey PJ (1980) J Chem Soc Perkin Trans 2:1313
67. De Lijser HJP, Rangel NA, Tetalman MA, Tsai C-K (2007) J Org Chem 72:4126
68. Schneider H-J, Sauerbrey R (1987) J Chem Res 14

69. Streith J, Danner B, Sigwalt C (1967) J Chem Soc Chem Commun 979
70. Boberg F, Steigerwald E (1979) Erdöl Kohle Erdgas Petrochem 569
71. Ogawa Y, Iwasaki S, Okuda S (1981) Tetrahedron Lett 22:2277
72. Akhtar MN, Boyd DR, Neill JD, Jerina DM (1980) J Chem Soc Perkin Trans 1:1693
73. Bhardwaj RK, Davidson RS (1987) J Chem Res 406
74. Decout J-L, LaBlache-Combier A, Loucheux C (1980) J Polym Sci Polym Chem 18:2371
75. Sammes PG, Serra-Errante G, Tinker AC (1978) J Chem Soc Perkin Trans 1:853
76. Serra-Errante G, Sammes PG (1975) J Chem Soc Chem Commun 574
77. Kubota T, Yamakawa M, Mizuno Y (1972) Bull Chem Soc Jpn 45:3282
78. Yamakawa M, Ezumi K, Mizuno Y, Kubota K (1974) Bull Chem Soc Jpn 47:2982
79. Bucher G, Scaiano JC (1994) J Phys Chem 98:12471
80. Scaiano JC, Bucher G, Barra M, Weldon D, Sinta R (1996) J Photochem Photobiol A Chem 102:7
81. Albini A, Fasani E, Frattini V (1987) J Photochem 37:355
82. Bellamy F, Barragan LGR, Streith J (1971) J Chem Soc Chem Commun 456
83. Carraher JM, Bakac A (2014) Phys Chem Chem Phys 16:19429
84. Geletii YV, Zubarev VE, Levin PP, Lyubimova GV, Shafirovich VY (1991) Kinetika I Kataliz (Engl Trans) 31:701
85. Geletii YV, Strelets VV, Shafirovich VY, Shilov AE (1989) Heterocycles 2:677
86. Nakanishi I, Nishizawa C, Ohkubo K, Takeshita K, Suzuki KT, Ozawa T, Hecht S, Tanno M, Sueyoshi S, Miyata N, Okuda H, Fukuzumi S, Ikota N, Fukuhara K (2005) Org Biomol Chem 3:3263
87. Wang L, Yang X, Zhao J, Zhang F, Wang X, Sun L (2014) ChemSusChem 7:2640
88. Tsuchiya T, Arai H, Igeta H (1972) J Chem Soc Chem Commun 550
89. Tsuchiya T, Arai H, Igeta H (1973) Chem Pharm Bull 21:1516
90. Tsuchiya T, Arai H, Igeta H (1973) Tetrahedron 29:2747
91. Kumler PL, Buchardt O (1968) J Am Chem Soc 90:5640
92. Portillo M, Maxwell MA, Frederich JH (2016) Org Lett 18:5142
93. Tsuchiya T, Arai H, Igeta H (1969) Tetrahedron Lett 32:2747
94. Bellamy F, Martz P, Streith J (1974) Tetrahedron Lett 36:3189
95. Bose S, Kumar S, Davies RJH, Sethi S, McCloskey JA (1984) J Chem Soc Perkin Trans 1:2421
96. Lam FL, Brown GB, Parham JC (1974) J Org Chem 39:1391
97. Lam FL, Parham JC (1975) J Am Chem Soc 97:2839
98. Lam FL, Parham JC (1973) J Org Chem 38:2397
99. Giner-Sorolla A, Gryte C, Cox ML, Parham JC (1971) J Org Chem 36:1228
100. Lam FL, Parham JC (1982) Tetrahedron 38:2371
101. Streith J, Darrah HK, Weil M (1966) Tetrahedron Lett 45:5555
102. Albini A, Fasani E, Dacrema LM (1980) J Chem Soc Perkin Trans 1:2738
103. Somei M, Kitamura R, Fujii H, Hashiba K, Kawai S, Kaneko C (1977) J Chem Soc Chem Commun 899
104. Kaneko C, Fujii H, Kawai S, Hashiba K, Karasawa Y, Wakai M, Hayashi R, Somei M (1982) Chem Pharm Bull 30:74
105. Glasnov TN, Kappe CO (2007) QSAR Comb Sci 26:1261
106. Fielden R, Meth-Cohn O, Suschitzky H (1973) J Chem Soc Perkin Trans 1:702
107. Solntsev KM, Clower CE, Tolbert LM, Huppert D (2005) J Am Chem Soc 127:8534
108. Albini A, Bettinetti GF, Minoli G (1979) Tetrahedron Lett 39:3761
109. Ishikawa M, Yamada S, Hotta H, Kaneko C (1966) Chem Pharm Bull 14:1102
110. Ishikawa M, Yamada S, Kaneko C (1965) Chem Pharm Bull 13:747
111. Kaneko C, Kitamura R (1977) Heterocycles 6:111
112. Kaneko C, Hayashi S, Kobayashi Y (1974) Chem Pharm Bull 22:2147
113. Kaneko C, Kitamura R (1977) Heterocycles 6:117
114. Kaneko C, Yokoe I, Yamada S, Ishikawa M (1969) Chem Pharm Bull 17:1290

115. Lei Y-J, Zhu C-H, Wang Z-Q, Liu H (2011) Chem Res Chin Univ 17:69
116. Kaneko C, Hasegawa H, Tanaka S, Sunayashaki K, Yamada S (1974) Chem Lett 133
117. Kasama K, Takematsu A, Yamamoto S, Arai S (1984) J Phys Chem 88:4918
118. Seki H, Takematsu A, Arai S (1987) J Phys Chem 91:176
119. Sheng Z, Song Q, Gao F, Zhou X, Li J, Dai J, Sun H, Li Q, Yu S, Ma X (2000) Res Chem Intermed 26:715
120. Choudhury S, Basu S (2006) J Phys Chem B 110:8850
121. Liu JF, Li X-Y, Zhu Q (2004) Int J Quantum Chem 98:33
122. Shi X, Platz MS (2004) J Phys Chem A 108:4385
123. Kurasawa Y, Takada A, Kim HS (1995) J Heterocycl Chem 32:1085
124. Albini A, Colombi R, Minoli G (1978) J Chem Soc Perkin Trans 1:924
125. Albini A, Fasani E, Frattini V (1988) J Chem Soc Perkin Trans 2:235
126. Tokumura K, Goto H, Kashiwabara H, Kaneko C, Itoh M (1980) J Am Chem Soc 102:5643
127. Kaneko C, Hayashi R, Yamamori M, Tokumura K, Itoh M (1978) Chem Pharm Bull 26:2508
128. Tokumura K, Matsushita Y (2001) J Photochem Photobiol A Chem 140:27
129. Ishikawa M, Kaneko C, Yamada S (1968) Tetrahedron Lett 43:4519
130. Kaneko C, Yamada S, Ishikawa M (1969) Chem Pharm Bull 17:1294
131. Yamada S, Ishikawa M, Kaneko C (1972) Tetrahedron Lett 11:971
132. Yamada S, Ishikawa M, Kaneko C (1975) Chem Pharm Bull 23:2818
133. Kaneko C, Okuda W, Karasawa Y, Somei M (1980) Chem Lett 547
134. Girreser U, Hermann TW, Piotrowska K (2003) Arch Pharm Pharm Med Chem 336:401
135. Suau R, Rico-Gomez R, Souto-Bachiller FA, Rodriguez-Rodriguez L, de Los A, Ruiz ML (1995) Tetrahedron Lett 36:2653
136. Souto-Bachiller FA, Perez-Inestrosa E, Suau R, Rico-Gomez R, Rodriguez-Rodriguez LA, Coronado-Pérez ME (1999) Photochem Photobiol 70:875
137. Collado D, Perez-Inestrosa E, Suau R (2003) J Org Chem 68:3574
138. Collado D, Perez-Inestrosa E, Suau R, Navarrete JTL (2006) Tetrahedron 62:2927
139. Lam FL, Lee T-C (1978) J Org Chem 43:167
140. Sako M, Ohara S, Hirota K, Maki Y (1990) Tetrahedron 46:4171
141. Sako M, Shimada K, Hirota K, Maki Y (1986) Tetrahedron Lett 27:3877
142. Sako M, Ohara S, Hirota K, Maki Y (1990) J Chem Soc Perkin Trans 1:3339
143. Sako M, Shimada K, Hirota K, Maki Y (1985) Tetrahedron Lett 26:6493
144. Sako M, Shimada K, Hirota K, Maki Y (1986) J Chem Soc Chem Commun 1705
145. Maki Y, Shimada K, Sako M, Hirota K (1988) Tetrahedron 44:3187
146. Sako M, Shimada K, Hirota K, Maki Y (1986) J Am Chem Soc 108:6039
147. Sako M, Takeda Y, Hirota K, Maki Y (1996) Heterocycles 42:31
148. Lin S-K, Feng L-B (1986) Chem Phys Lett 128:319
149. Wang H, Feng L (1988) J Phys Org Chem 1:191
150. Lin S-K (1987) J Photochem 37:363
151. Lin S, Wang H (1986) Heterocycles 24:659
152. Kawata H, Niizuma S, Kokubun H (1980) J Photochem 13:261
153. Arai H, Ohsawa A, Saiki K, Igeta H (1977) J Chem Soc Chem Commun 133
154. Ohsawa A, Arai H, Igeta H, Akimoto T, Tsuji A (1979) Tetrahedron 35:1267
155. Ershov Y, Kovalenko NP, Orlova SF, Shekk Yu B (1986) Izvest Akademii Nauk SSSR Ser Khim (Engl Transl) 806
156. De Vries H, Bojarski J, Donker AA, Bakri A, Beyersbergen van Henegouwen GMJ (1990) Toxicology 63:85
157. Haddadin MJ, Issidorides CH (1993) Heterocycles 35:1503
158. Haddadin MJ, Agopian G, Issidorides CH (1971) J Org Chem 36:514
159. Masoud NA, Olmsted J (1975) J Phys Chem 79:2214
160. Jarrar AA, Halawi SS, Haddadin MJ (1976) Heterocycles 4:1077
161. Jarrar AA (1978) J Heterocycl Chem 15:177

162. Baas P, Oppelaar H, Stavenuiter M, van Zandwijk N, Stewart FA (1993) Int J Radiat Oncol Biol Phys 27:665
163. Adams GE, Stratford IJ (1994) Int J Radiat Oncol Biol Phys 29:231
164. Shi X, Poole JS, Emenike I, Burdzinski G, Platz MS (2005) J Phys Chem A 109:1491
165. Laderoute K, Wardman P, Rauth AM (1988) Biochem Pharmacol 37:1487
166. Poole JS, Hadad CM, Platz MS, Fredin ZP, Pickard L, Levya Guerrero E, Kessler M, Chowdhury G, Kotandeniya D, Gates KS (2002) Photochem Photobiol 75:339
167. Daniels JS, Chatterji T, MacGillivray LR, Gates KS (1998) J Org Chem 63:10027
168. Kawata H, Niizuma S, Kumagi T, Kokubun H (1985) Chem Lett 767
169. Kawata H, Niizuma S, Kokubun H (1978) J Photochem 9:463
170. Wosinska ZM, Stump FL, Ranjan R, Lorance ED, Finley GN, Patel PP, Khawaja MA, Odom KL, Kramer WH, Gould IR (2014) Photochem Photobiol 90:313
171. Hamana M, Noda H (1969) Chem Pharm Bull 17:2633
172. Huntley JJA, Nieman R, Rose SD (1999) Photochem Photobiol 69:1
173. Collado D, Perez-Inestrosa E (2012) Eur Polym J 8100
174. Mee JD, Heseltine DW, Taylor EC (1970) J Am Chem Soc 92:5814
175. Shukla D, Liu G, Dinnocenzo JP, Farid S (2003) Can J Chem 81:744
176. Baciocchi E, Del Giacco T, Gerini MF, Lanzalunga O (2006) Org Lett 8:641
177. Del Giacco T, Lanzalunga O, Mazzonna M, Mencarelli P (2012) J Org Chem 77:1843
178. Bettoni M, Del Giacco T, Stradiotto M, Fausto E (2015) J Org Chem 80:8001
179. Barbieri A, Chimenti RD, Del Giacca T, Di Stefano S, Lanzalunga O, Lapi A, Mazzona M, Olivo G, Salamone M (2016) J Org Chem 81:2513
180. Lee KY, Kochi JK (1992) J Chem Soc Perkin Trans 2:1011
181. Bockman TM, Lee KY, Kochi JK (1992) J Chem Soc Perkin Trans 2:1581
182. Yildirim TG, Hepuzer Y, Hizal G, Yagci Y (1999) Polymer 40:3885
183. Hizal G, Emiroglu SE, Yagci Y (1998) Polym Int 47:391
184. Gould IR, Shukla D, Giesen D, Farid S (2001) Helv Chim Acta 84:2796
185. Podsiadly R (2008) J Photochem Photobiol A Chem 198:60
186. Kolinska J, Podsiadly R, Wysocki S, Sokolowska J (2009) Dyes Pigments 82:238
187. Podsiadly R (2009) J Photochem Photobiol A Chem 202:115
188. Podemska K, Podsiadly R, Szymczak AM, Sokolowska J (2012) Dyes Pigments 94:113
189. Podemska K, Podsiadly R, Orzel A, Kowalska A, Maruszewska A, Marcinek A, Sokolowska J (2014) Color Technol 130:250
190. Lorance ED, Kramer WH, Gould IR (2002) J Am Chem Soc 124:15225
191. Lorance ED, Hendrickson K, Gould IR (2005) J Org Chem 70:2014
192. Lorance ED, Kramer WH, Gould IR (2004) J Am Chem Soc 126:14071
193. Kabatc J, Paczkowski J (2005) Macromolecules 38:9985
194. Kabatc J, Paczkowski J (2010) J Appl Polym Sci 117:2669
195. Kabatc J, Paczkowski J (2006) Polymer 47:2699
196. Kabatc J, Jedrzejewska B, Packowski J (2006) Macromol Mater Eng 291:646
197. Shukla D, Ahern WG, Farid S (2005) J Org Chem 70:6809
198. Shukla D, Ahearn WG, Farid S (2006) Photochem Photobiol 82:146
199. Shukla D, Adiga SP, Ahern WG, Dinnocenzo JP, Farid S (2013) J Org Chem 78:1955
200. Quint V, Morlet-Savary F, Lohier J-F, LaLevée J, Gaumont A-C, Lakhdar S (2016) J Am Chem Soc 138:7436
201. Zhu QQ, Schnabel W (1996) Polymer 37:4129
202. Acar MH, Yagci Y, Schnabel W (1998) Polym Int 46:331
203. Atmaca L, Onen A, Yagci Y (2001) Eur Polym J 37:677
204. Bacak V, Reetz I, Yagci Y, Schnabel W (1998) Polym Int 47:345
205. Hepuzer Y, Kücüktönbekici U, Yagci Y (2000) J Photochem Photobiol A Chem 130:71
206. Onen A, Yagci Y (2001) Macromol Chem Phys 202:1950
207. Monecke P, Schnabel W, Yagci Y (1997) Polymer 38:5389
208. Reetz I, Bacak V, Yagci Y (1997) Polymer Int 43:27

209. Degirmenci M, Onen A, Yagci Y, Pappas SP (2001) Polym Bull 46:443
210. Degirmenci M, Hepuzer Y, Yagci Y (2002) J Appl Polym Sci 85:2389
211. Dursun C, Degirmenci M, Yagci Y, Jockusch S, Turro NJ (2003) Polymer 44:7389
212. Arsu N, Önen A, Yagci Y (1996) Macromolecules 29:8973
213. Titskii GD, Turovskaya MK (2002) Theor Exp Chem 38:192
214. Sakurai T, Inoue H (1984) J Chem Soc Perkin Trans 2:2031
215. Aveline BM, Kochevar IE, Redmond RW (1996) J Am Chem Soc 118:10124
216. Adam W, Marquardt S, Saha-Möller CR (1999) Photochem Photobiol 70:287
217. Nazarova AI, Feofanov AV, Karmakova TA, Sharonov GV, Plyutinskaya AD, Yakubovskaya RI, Lebedeva VS, Mironov AF, Maurizot J-C, Vigny P (2005) Russ J Biorg Chem 31:482
218. Sakurai T, Murakata Y, Inoue H (1990) J Chem Soc Perkin Trans 2:499
219. Adam W, Hartung J, Okamoto H, Marquardt S, Nau WM, Pischel U, Saha-Möller CR, Spehar K (2002) J Org Chem 67:6041
220. Adam W, Marquardt S, Kemmer D, Saha-Möller CR, Schreier P (2002) Org Lett 4:225
221. Adam W, Marquardt S, Kemmer D, Saha-Möller CR, Schreier P (2002) Photochem Photobiol Sci 1:609
222. Tanaka K, Nakamura K, Yoshioka N, Kameyama A, Igarashi T, Sakuai T (2004) J Polym Sci A Polym Chem 42:2859
223. Yoshioka N, Andoh C, Kubo K, Igarashi T, Sakurai T (2001) J Chem Soc Perkin Trans 2:1927
224. Furrer H (1974) Tetrahedron Lett 34:2953
225. Katritzky AR, Chapman AV, Cook MJ, Millet GH (1979) J Chem Soc Chem Commun 396
226. Katritzky AR, Chapman AV, Cook MJ, Millet GH (1980) J Chem Soc Perkin Trans 1:2743
227. Taylor EC, Altland HW, Kienzle F, McKillop A (1976) J Org Chem 41:24
228. Ohkanda J, Kumasaka T, Takasu A, Hasegawa T, Katoh A (1996) Heterocycles 43:883
229. Parham JC, Pullman I, Brown GB (1973) Tetrahedron 29:3329
230. Parham JC, Templeton MA (1980) Tetrahedron 36:709
231. Aveline BM, Kochevar I, Redmond RW (1996) J Am Chem Soc 118:10113
232. Botchway SW, Crisostomo AG, Parker AW, Bisby RH (2007) Arch Biochem Biophys 464:314
233. DeMatteo MP, Poole JS, Shi X, Sachdeva R, Hatcher PG, Hadad CM, Platz MS (2005) J Am Chem Soc 127:7094
234. Poole JS, Shi X, Hadad CM, Platz MS (2005) J Phys Chem A 109:2547
235. Mitroka S, Zimmeck S, Troya D, Tanko JM (2010) J Am Chem Soc 132:2907
236. Marin ML, Lhiaubet-Vallet V, Santos-Juanes L, Soler J, Gomis J, Arques A, Amat AM, Miranda MA (2011) Appl Catal Environ 103:48
237. Aveline BM, Kochevar IE, Redmond RW (1996) J Am Chem Soc 118:289
238. Reszka KJ, Chignell CF (1994) Photochem Photobiol 60:442
239. Reszka KJ, Chignell CF (1995) Photochem Photobiol 61:269
240. Fukuda R, Ehara M (2016) Theor Chem Acc 135:105
241. Alam MM, Watanabe A, Ito O (1996) Photochem Photobiol 63:53
242. Adam W, Ballmeier D, Epe B, Grimm GN, Saha-Möller CR (1995) Angew Chem Intl Edn Engl 34:2156
243. Chaulk SG, Pezacki JP, MacMillan AM (2000) Biochemistry 39:10448
244. Douki T, Spinelli S, Ravanat J-L, Cadet J (1999) J Chem Soc Perkin Trans 2:1875
245. Telo JP (2003) Org Biomol Chem 1:588
246. Vrantza D, Kaloudis P, Liondiadis L, Gimisis T, Vougioukalakis GC, Orfanopoulos M, Gasparutto D, Cadet J, Encinas S, Paris C, Miranda MA (2006) Helv Chim Acta 89:2371
247. Sakkas VA, Shibata K, Yamaguchi Y, Sugasawa S, Albanis T (2007) J Chromatog A 1144:175
248. Aveline BM, Redmond RW (1999) J Am Chem Soc 121:9977
249. Adam W, Grimm GN, Marquardt S, Saha-Möller CR (1999) J Am Chem Soc 121:1179

250. Möller M, Adam W, Marquardt S, Saha-Möller CR, Stopper H (2005) Free Radic Biol Med 39:473
251. Arnone M, Hartung J, Engels B (2005) J Phys Chem A 109:5943
252. Arnone M, Engels B (2007) J Phys Chem A 111:3161
253. Arnone M, Engels B (2006) J Phys Chem A 110:12330
254. Hartung J, Hiller M, Schmidt P (1996) Chem A Eur J 2:1014
255. Hartung J, Hiller M, Schmidt P (1996) Liebigs Ann 1425
256. Hartung J, Kneuer R (2000) Eur J Org Chem 1677
257. Hartung J, Kneuer R, Kopf TM, Schmidt P (2001) CR Acad Sci Paris Chemie 4:649
258. Hartung J, Kempter I, Gottwald T, Schwarz M, Knauer R (2009) Tetr Asymm 20:2097
259. Ko EJ, Williams CM, Savage GP, Tsanaktsidis J (2012) Organic Synth 89:471
260. Aveline BM, Kochevar IE, Redmond RW (1995) J Am Chem Soc 117:9699
261. Barton DHR, Bridon D, Fernandez-Picot I, Zard SZ (1987) Tetrahedron 43:2733
262. Barton DHR, Hervé Y, Potier P, Thierry J (1987) Tetrahedron 43:4297
263. Castagnino E, Corsano S, Barton DHR (1989) Tetrahedron Lett 30:2983
264. Barton DHR, Crich D, Kretzschmar G (1986) J Chem Soc Perkin Trans 1:39
265. Barton DHR, Gateau-Olesker A, Gero SD, Lacher B, Tachdijan C, Zard SZ (1987) J Chem Soc Chem Commun 1790
266. Barton DHR, da Silva E, Zard SZ (1988) J Chem Soc Chem Commun 285
267. Barton DHR, Gero SD, Quiclet-Sire B, Samadi M (1988) J Chem Soc Chem Commun 1372
268. Barton DHR, Ozbalik N, Vacher B (1988) Tetrahedron 44:7385
269. Barton DHR, Garcia B, Togo H, Zard SZ (1986) Tetrahedron Lett 27:1327
270. Barton DHR, Tachdijan C (1992) Tetrahedron 48:7109
271. Barton DHR, Chern C-Y, Jaszberenyi JC (1992) Tetrahedron Lett 35:5017
272. Barton DHR, Ozbalik N, Vacher B (1988) Tetrahedron 44:3501
273. Barton DHR, Ramesh M (1990) Tetrahedron Lett 31:949
274. Lapinski L, Gerega A, Sobolewski AL, Nowak MJ (2008) J Phys Chem A 112:238
275. Barton DHR, Blundell P, Jaszberenyi JC (1991) J Am Chem Soc 113:6937
276. Bohne C, Boch R, Scaiano JC (1990) J Org Chem 55:5414
277. Schepp NP, Green CJ, Cozens FL (2010) Photochem Photobiol Sci 9:110
278. Horner JH, Tanaka N, Newcomb M (1998) J Am Chem Soc 120:10379
279. Emanuel CJ, Horner JH, Newcomb M (2000) J Phys Org Chem 13:688
280. DeZutter CB, Horner JH, Newcomb M (2008) J Phys Chem A 112:1891
281. DeZutter CB, Horner JH, Newcomb M (2011) Org Biomol Chem 9:516
282. Bales BC, Horner JH, Huang X, Newcomb M, Crich D, Greenberg MM (2001) J Am Chem Soc 123:3623
283. Newcomb M, Miranda N, Huang X, Crich D (2000) J Am Chem Soc 122:6128
284. Horner JH, Taxil E, Newcomb M (2002) J Am Chem Soc 124:5402
285. Taxil E, Bagnol L, Horner JH, Newcomb M (2003) Org Lett 5:827
286. Broyles DA, Carpenter BK (2005) Org Biomol Chem 3:1757
287. Kawamoto T, Uehara S, Hirao H, Fukuyama T, Matsubara H, Ryu I (2014) J Org Chem 79:3999
288. He W, Togo H, Ogawa H, Yokoyama M (1997) Heteroatom Chem 8:411
289. Barton DHR, Castagnino E, Jaszberenyi JC (1994) Tetrahedron Lett 35:6057
290. Dauben WG, Kowalczyk BA, Bridon DP (1989) Tetrahedron Lett 30:2461
291. Eaton PE, Maggini M (1988) J Am Chem Soc 110:7230
292. Theodorakis EA, Xiang X, Blom P (1997) Chem Commun 1463
293. Castagnino E, Corsano S, Barbaro R, Carloni P (1998) J Chem Res 384
294. Castagnino E, Cangiotti M, Tongianai S, Ottaviani MF (2005) J Control Release 108:215
295. Aveline BM, Redmond RW (1998) Photochem Photobiol 68:266
296. Castagnino E, Ottavani MF, Cangiotti M, Morelli M, Casettari L, Muzzarelli RAA (2008) Carbohydr Polym 74:640
297. Ling T, Poupon E, Rueden EJ, Kim SH, Theodorakis EA (2002) J Am Chem Soc 124:12261

298. Okano T, Ishihara H, Takakura N, Tsuge H, Eguchi S, Kimoto H (1997) J Am Chem Soc 62:7192
299. Ziegler FE, Wang Y (1996) Tetrahedron Lett 35:6299
300. Rochelle N, Cormier T, Ahmed-Schofield R (2006) J Undergrad Res 3:131
301. Cormack PAG, Shirshova N, Steinke JHG (2005) Ind Eng Chem Res 44:8699
302. Newcomb M, Kumar MU (1991) Tetrahedron Lett 32:45
303. Togo Y, Nakamura N, Iwamura H (1991) Chem Lett 1201
304. Schultz BE, Hansen KC, Lin CC, Chan SI (2000) J Org Chem 65:3244
305. Newcomb M, Deeb TM (1987) J Am Chem Soc 109:3163
306. Esker JL, Newcomb M (1992) Tetrahedron Lett 33:5913
307. Burdi D, Aveline BM, Wood PD, Stubbe J, Redmond RW (1997) J Am Chem Soc 119:6457
308. Schiesser CH, Sutej K (1992) J Chem Soc Chem Commun 57
309. Fong MC, Schiesser CH (1993) Tetrahedron Lett 34:4347
310. Liebscher J, Riemer B, Bendig J, Stösser R (1994) Tetrahedron Lett 35:7009

Index

MIX
Papier aus verantwortungsvollen Quellen
Paper from responsible sources
FSC® C105338

If you have any concerns about our products, you can contact us on
ProductSafety@springernature.com

In case Publisher is established outside the EU, the EU authorized representative is:
Springer Nature Customer Service Center GmbH
Europaplatz 3, 69115 Heidelberg, Germany

Printed by Libri Plureos GmbH
in Hamburg, Germany